Family Math Night K-5

Host family math nights at your elementary school—starting today! Family math nights are a great way for teachers to get parents involved in their children's education and to promote math learning outside of the classroom. In this practical book, you'll find step-by-step guidelines and activities to help you bring family math nights to life. The enhanced second edition is aligned with the Common Core State Standards for Mathematical Content and Practice with new activities to help students explain their answers and write about math. It also comes with ready-to-use handouts that you can distribute during your event. With the resources in this book, you'll have everything you need to help students learn essential math concepts—including counting and cardinality; operations and algebraic thinking; number and operations in base ten; number and operations—fractions; measurement and data; and geometry—in a fun and supportive environment.

Special Features:

- The book is organized by math content and grade band, so you can quickly find activities that meet your needs.

- Each activity is easy to implement and includes a page of instructions educators can use to prepare the station, as well as a page for families that explains the activity and can be photocopied and displayed at the station.

- All of the family activities can be photocopied or downloaded from our website, www.routledge.com/9781138915541, so that you can distribute them during your event.

Jennifer Taylor-Cox is owner of Taylor-Cox Instruction, LLC. She serves as an educational consultant for numerous districts across the United States. She teaches university courses in education and is author of seven Routledge Eye On Education books.

Also available from Routledge Eye On Education
(www.routledge.com/eyeoneducation)

Family Math Night 6–8, Second Edition
Math Standards in Action
Jennifer Taylor-Cox and Christine Oberdorf

Math Intervention 3–5, Second Edition
Building Number Power with Formative Assessments,
Differentiation, and Games
Jennifer Taylor-Cox

Math Intervention PreK-2, Second Edition
Building Number Power with Formative Assessments,
Differentiation, and Games
Jennifer Taylor-Cox

Using Formative Assessment to Drive Mathematics Instruction in Grades 3–5
Jennifer Taylor-Cox and Christine Oberdorf

Using Formative Assessment to Drive Mathematics Instruction in Grades PreK-2
Jennifer Taylor-Cox and Christine Oberdorf

Solving Behavior Problems in Math Class
Academic, Learning, Social, and Emotional Empowerment,
Grades K-12
Jennifer Taylor-Cox

Family Math Night K-5

Common Core State Standards in Action

Second Edition

Jennifer Taylor-Cox

Routledge
Taylor & Francis Group
NEW YORK AND LONDON

Second edition published 2018
by Routledge
711 Third Avenue, New York, NY 10017

and by Routledge
2 Park Square, Milton Park, Abingdon, Oxon, OX14 4RN

Routledge is an imprint of the Taylor & Francis Group, an informa business

© 2018 Taylor & Francis

The right of Jennifer Taylor-Cox to be identified as author of this work has been asserted by her in accordance with sections 77 and 78 of the Copyright, Designs and Patents Act 1988.

All rights reserved. No part of this book may be reprinted or reproduced or utilized in any form or by any electronic, mechanical, or other means, now known or hereafter invented, including photocopying and recording, or in any information storage or retrieval system, without permission in writing from the publishers.

Trademark notice: Product or corporate names may be trademarks or registered trademarks, and are used only for identification and explanation without intent to infringe.

First edition published by Eye On Education 2005

Library of Congress Cataloging-in-Publication Data
A catalog record has been requested for this book

ISBN: 978-1-138-91553-4 (hbk)
ISBN: 978-1-138-91554-1 (pbk)
ISBN: 978-1-315-69016-2 (ebk)

Typeset in Times New Roman
by Florence Production Ltd, Stoodleigh, Devon, UK

Art by Maria Diaz Cassi
Visit the companion website: www.routledge.com/9781138915541

About the Author

Dr. Jennifer Taylor-Cox is an energetic, captivating presenter and well-known educator. She is the owner of Taylor-Cox Instruction: Connecting Research and Practice in Education.

Jennifer serves as an educational consultant for numerous districts across the United States. Her workshops and keynote speeches are always high-energy and insightful. She earned her Ph.D. from the University of Maryland and was awarded the "Outstanding Doctoral Research Award" from the University of Maryland and the "Excellence in Teacher Education Award" from Towson University. She currently serves as the president-elect of the Maryland Council of Teachers of Mathematics. Jennifer truly understands how to connect research and practice in education. Her passion for mathematics education is alive in her work with students, parents, teachers, and administrators.

Dr. Taylor-Cox lives and has her office in Severna Park, Maryland. She is the mother of three children.

If you would like to have Dr. Taylor-Cox present a Family Math Night at your school or if you would like to schedule professional development opportunities for educators and/or parents, please contact her at Taylor-Cox Instruction: Connecting Research and Practice in Education.

Jennifer Taylor-Cox, Ph.D., Educational Consultant
Office: 410-729-5599, Fax: 410-729-3211
Email: jennifer@taylor-coxinstruction.com

About Taylor-Cox Instruction

Taylor-Cox Instruction provides professional development for Pre-K through Grade 12 educators. Jennifer Taylor-Cox and her associates will work with you to design precise and effective professional development opportunities for the educators in your school, district, state, or region. Each professional development opportunity is catered to meet the specific needs of your students, educators, and parents.

Differentiating Math Instruction: Target instruction to meet the learning needs of all students

Math Intervention: Help struggling ELLs, special needs, and other students find success

Classroom Discipline: Meet the challenges and help all students succeed

English Language Learners: Increase content discourse and conceptual understanding for ELL students

Building Number Power: Bolster all students' number sense and computation

Family Math Nights: Involve and engage parents and students in learning

Sigmund Square Finds His Family
Published by Taylor-Cox Instruction

Free download of the interactive eBook available at www.routledge.com/9781138915541
To purchase a traditional paperback (978-0-9838880-0-0) or PDF (978-0-9838880-1-7), visit www.sigmundsquare.com

jennifer@taylor-coxinstruction.com · www.taylor-coxinstruction.com · www.sigmundsquare.com

Meet the Illustrator

It is a pleasure to introduce Maria Diaz Cassi, the illustrator of *Family Math Night*. I met Maria nearly 30 years ago when she walked into the classroom on the first day of school. I soon learned that she was a smart, creative, compassionate, and determined little first grader. I loved teaching Maria mathematics, reading, writing, science, and social studies. She expressed such a great joy for learning. She loved to draw and doodle. She shared her thoughts of justice and fairness. She advocated for others. It was an honor to be the teacher of such an incredible young lady.

How wonderful it was that we ran into each other a couple of years ago! As a teacher, it is a rare opportunity to see your students much later in their lives. I was not surprised to learn that Maria is still smart, creative, compassionate, and determined. She still believes in justice and serves as advocate for many. Her drawings and doodles have become professional, and I am proud that she created the illustrations for this book.

Our favorite poem in first grade, "Hug O' War" by Shel Silverstein, really describes the world to which we both aspire:

> I will not play at tug o' war.
> I'd rather play at hug o' war,
> Where everyone hugs
> Instead of tugs,
> Where everyone giggles
> And rolls on the rug,
> And everyone kisses,
> And everyone grins,
> And everyone cuddles,
> And everyone wins.

—Jennifer Taylor-Cox

From Maria:

Anytime anyone asks me, "Who was your favorite teacher and why?" my answer has always been the same, "Dr. Taylor-Cox." She was and still is an amazing teacher. Dr. Taylor-Cox has always been able to make learning fun, rather than burdensome. My memories of being in her classroom, nearly 30 years ago, are filled with smiles and games.

It was a huge honor to work with her and be a part of this project. While working with her, I remembered why I enjoyed being in her class so much. She continues to motivate and spark a desire to learn in others. She is still the same kind, caring, and warm person that she was 30 years ago.

I am so happy that I was able to work with her on this book as the illustrator. It is not every day that you can say that you were not only able to be reunited with your first grade teacher, but able to work on a book with her too!

—*Maria Diaz Cassi*

eResources

All of the family pages in this book are available as free downloads on our website, www.routledge.com/9781138915541, so you can easily print and use them during your events.

Contents

Chapter 1: Introduction — 1

    Why Should Our School Have Family Math Night? — 2

    How Is the Book Organized? — 3

    How Are the Activities Connected to the Common Core State Standards? — 3

    Why Should We Use Manipulatives in Mathematics? — 5

    Why Are Questions Included? — 5

    Why Is There a Challenge for Each Activity? — 6

    What Are Some Additional Tips? — 6

    What Is the Teacher's Role During Family Math Night? — 7

Chapter 2: Primary Stations:
Prekindergarten Through Grade 2 — 9

    Counting and Cardinality — 10
- Count and Ring the Bell — 10
- Stop and Go Counting — 12
- Is It More or Less? — 14
- Cereal Chains — 16

    Operations and Algebraic Thinking — 18
- Let's Shake It Up — 18
- Frog Hop Addition — 20
- Lu-Lu — 22
- Magic Squares — 24

    Number and Operations in Base Ten — 26
- Hundred Chart Tic-Tac-Toe — 26
- Pinch and Spoon — 28
- Ten Count — 30
- Match and Take — 32

    Measurement and Data — 34
- My Shoes — 34
- Bubbles Bubbles — 36
- Math Pong — 38
- Race to Midnight — 40

Geometry — 42
- Geoboard Shapes — 42
- Tangrams — 44
- Composing Hexagons — 46
- Sigmund Square — 48

Chapter 3: Intermediate Stations: *Grade 3 Through Grade 5* — 51

Operations and Algebraic Thinking — 52
- Snack Shop — 52
- Multiply on the Fly — 54
- Flip Top Division — 56
- Hooray Array — 58

Number and Operations in Base Ten — 60
- Place the Digit — 60
- Card Math — 62
- Palindromes — 64
- Heads Up Decimals — 66

Number and Operations—Fractions — 68
- Where's ½? — 68
- Clothesline Fractions — 70
- Fishing for Fractions — 72
- Bear Slide — 74

Measurement and Data — 76
- Caterpillar Collection — 76
- Honeybees — 78
- Layers of Cake — 80
- Angle Face — 82

Geometry — 84
- Draw What I Say — 84
- Pentominoes — 86
- Three Smiles — 88
- Geometry Page by Page — 90

Chapter 4: Resources — 93

Family Math Night Invitation to Parents — 95
Family Math Night Journal Cover — 96
Family Math Night Evaluation — 97

Number Cards	98
Stop and Go Signs	99
Let's Shake It Up	100
Lu-Lu Chips	101
Hundred Chart	102
Pinch and Spoon Game Board	103
Ten Count Game Board	104
Match and Take Number Lines	105
Match and Take Expression Cards	106
Math Pong Bar Graph Template	107
Race to Midnight Time Gone By Cards	108
Race to Midnight Recording Sheet	109
Tangrams Answer Key	110
Composing Hexagons 1	111
Composing Hexagons 2	112
Pattern Blocks Name Chart	113
Sigmund Square Question Cards P–2	114
Sigmund Square Question Cards 3–5	115
Sigmund Square Game Board	116
Snack Shop Cards	117
Flip Top Division Bottle Cap Labels	118
Place the Digit Game Boards	119
Card Math Game Board	123
Heads Up Decimal Cards	124
Heads Up Decimal Grid	125
Where's $\frac{1}{2}$? Number Lines	126
Fraction Fish	129
Bear Slide Results Cards	130
Bear Slide Number Lines	131
Caterpillars	132
Angle Face Cards	133

Angle Face Sorting Mats	**134**
Draw What I Say Cards	**135**
Pentominoes	**136**
Three Smiles Coordinate Plane	**137**
Geometry Vocabulary	**138**

Chapter 1
Introduction

- Why Should Our School Have Family Math Night?

- How Is the Book Organized?

- How Are the Activities Connected to the Common Core State Standards?

- Why Should We Use Manipulatives in Mathematics?

- Why Are Questions Included?

- Why Is There a Challenge for Each Activity?

- What Are Some Additional Tips?

- What Is the Teacher's Role During Family Math Night?

Why Should Our School Have Family Math Night?

The goal of Family Math Night is to strengthen the mathematical aptitudes of students through the power of family interaction. By sponsoring Family Math Night, educators are encouraging parents and students to appreciate the energy and pleasure of mathematics. Each activity is designed to promote mathematical thinking and communication. The hands-on approach presented in this book helps make learning mathematics a meaningful and productive process for all involved.

Parents play an important role in the academic lives of students. By participating in Family Math Night, parents can serve as models of motivation, persistence, and competency to their children. The directions for each activity are presented in a clear, concise manner, allowing parents to guide students to more complete understandings of various mathematics concepts. At the same time, parents may be learning new knowledge and solidifying or revising previous knowledge about mathematics. You may hear parents saying, "I never really understood that concept until I tried this activity with my child," or "I never knew math could be so fun!" In many ways, the Family Math Night activities enlighten parents as they begin to understand and value mathematics in new ways.

The concepts presented in each Family Math Night activity will help students learn essential new skills and/or reinforce skills already learned in mathematics. While working through math problems in a textbook is one way for some students to learn mathematics, there are other more interactive means of gaining knowledge of mathematics, such as the math stations presented in this book. To help realize a vision of increased math proficiency for all, we need to encourage students to think about and apply mathematics in the real world. Family Math Night can help students become mathematically fulfilled and empowered!

How Is the Book Organized?

Family Math Night contains four chapters. The first chapter addresses the goals and intentions of this book. The second chapter presents 20 math stations for primary students (prekindergarten through second grade). The third chapter offers 20 math stations for intermediate students (third through fifth grades). The final chapter provides additional tools for the successful implementation of Family Math Night.

There are two pages for each math station. The first page offers a list of the materials needed for the station, helpful hints, and math standards in action. The second page offers the directions, questions to get students talking about math, and a challenge. The first page is for educators to review and use to prepare each station. The second page can be photocopied and displayed at the Family Math Night station. The directions page can be laminated and mounted. Some educators find it helpful to attach each direction sheet to a file folder. The opened folder can be placed vertically at each station. Other educators prefer to place the direction sheets into display stands or onto display boards. The point is to have the directions clearly displayed at each station.

How Are the Activities Connected to the Common Core State Standards?

The Common Core State Standards for Mathematics (2010) require deeper understanding of mathematics. These standards are based on rigorous content, application of knowledge, and alignment with college and career expectations.

The Common Core State Standards for Mathematical Content (2010) define what students should understand. Each Family Math Night station highlights a specific mathematical content domain.

Sometimes more than one domain is highlighted because mathematics concepts are related.

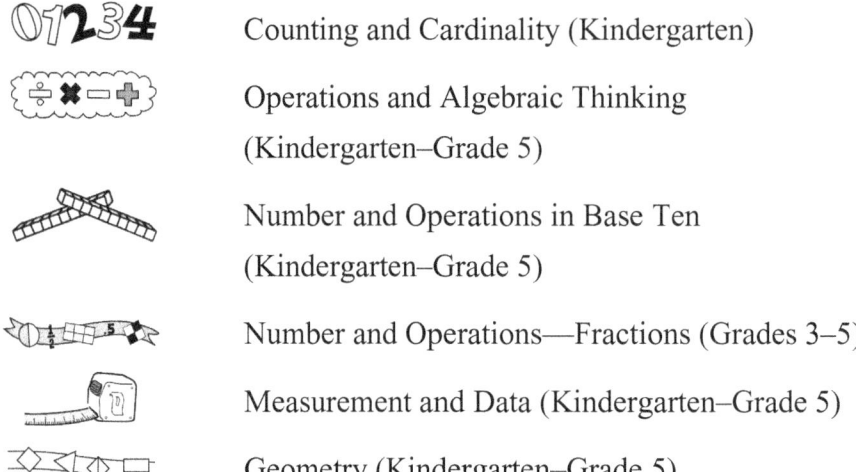

Counting and Cardinality (Kindergarten)

Operations and Algebraic Thinking (Kindergarten–Grade 5)

Number and Operations in Base Ten (Kindergarten–Grade 5)

Number and Operations—Fractions (Grades 3–5)

Measurement and Data (Kindergarten–Grade 5)

Geometry (Kindergarten–Grade 5)

The Common Core State Standards for Mathematical Practice (2010) describe what students should be doing while engaged in mathematics. Each Family Math Night station highlights two mathematical practices. This is not to imply that these are the only mathematical practices that are connected to the activity. The two mathematical practices listed for each activity were specifically considered during the design of the task and related questions.

- MP 1 Make sense of problems and persevere in solving them.
- MP 2 Reason abstractly and quantitatively.
- MP 3 Construct viable arguments and critique the reasoning of others.
- MP 4 Model with mathematics.
- MP 5 Use appropriate tools strategically.
- MP 6 Attend to precision.
- MP 7 Look for and make use of structure.
- MP 8 Look for and express regularity in repeated reasoning.

Why Should We Use Manipulatives in Mathematics?

Using manipulatives in mathematics allows students to experience abstract concepts in a concrete manner. Building models to represent math ideas and concepts strengthens the conceptual frameworks students construct as they apply math to everyday life. Manipulatives provide the means by which many students need to express the reasoning and evidence associated with the math thinking. Using manipulatives to show how one derives an answer helps solidify understanding. Manipulatives offer students the tools to solve mathematical situations. Additionally, manipulatives often serve as the springboard for math communication as students explain and justify how they solve a problem and/or approach a solution.

To encourage the successful use of math manipulatives, educators should think about how the manipulatives are organized and how they are made available to students. For example, sets of manipulatives can be prepared and stored in plastic bags, baskets, or other containers. The listed manipulatives need to be made available to students and parents at each Family Math Night station.

Why Are Questions Included?

Asking questions invites students to engage in mathematical communication. Questions promote mathematical thinking and encourage "math talk." We do not want the Family Math Night room to be a "quiet zone." Instead, we want to strive for a room full of active mathematics participants who are engaged in productive mathematics discourse. By promoting "math talk" at Family Math Night, we will better prepare our students for the mathematical challenges ahead. Our role is to provide students with opportunities to hear, use, and come to know the richness of "math talk."

Why Is There a Challenge for Each Activity?

Purposeful challenge serves to inspire and enlighten many students. Each Family Math Night activity includes a challenge that provides a possible extension of the activity. Sometimes students are so engaged in the activity that they want to investigate it further. Other times students go directly to the challenge as a way to increase the level of difficulty. Essentially, the challenges offer a way to differentiate the learning opportunities for children and their families.

What Are Some Additional Tips?

If you want high attendance at your Family Math Night, you need to advertise to students and the parents. Send home notices about the event (a sample notice to parents is found on page 95). Include the event in newsletters and on the school calendar. Offer incentives for students and parents, such as pizza or snacks. Some schools offer recognition to the class with the highest attendance. Other schools encourage students to attend by allowing participation in Family Math Night to serve as the night's homework. The possibilities are nearly endless!

To accommodate many families, you will need a large room or several large rooms. Position the tables in a manner that allows for maximum movement and comfort. Posting multiple copies of the directions and providing several sets of the materials allow you to have more than one family at each Family Math Night station. Also, color-coding the directions pages can be very helpful. For example, all of the primary activities can be photocopied on yellow paper and all of the intermediate stations can be photocopied on blue paper. This coding system helps parents direct their children toward grade-level appropriate activities. If you decide to color-code the activities, provide parents with a key indicating the coding system upon entry.

Providing a check-in table is a good idea. Parents can sign-in students or teachers can check-off students on class lists. The check-in area is a place where students can obtain pencils and Family Math Night journals. The journals can be simple booklets of blank pages for students to record information related to the activities. A sample Family Math Night journal cover is found on page 97. If you decide not to provide students with math journals, distribute paper and pencils. Evaluation forms can also be distributed at the check-in area. A sample Family Math Night evaluation form is found on page 96. The information gathered from the evaluations will help you plan subsequent Family Math Nights.

What Is the Teacher's Role During Family Math Night?

During Family Math Night, educators should highlight the mathematical endeavors of students and parents. While visiting families at each station, educators should also encourage math conversation and math thinking. Some of the consumable materials may need to be replenished, and some of the stations need to be monitored. However, be sure to take at least one moment during Family Math Night to notice how the event mathematically inspires and empowers students and parents!

Chapter 2
Primary Stations
Prekindergarten Through Grade 2

<u>Counting and Cardinality</u>
- Count and Ring the Bell
- Stop and Go Counting
- Is It More or Less?
- Cereal Chains

<u>Operations and Algebraic Thinking</u>
- Let's Shake It Up
- Frog Hop Addition
- Lu-Lu
- Magic Squares

<u>Number and Operations in Base Ten</u>
- Hundred Chart Tic-Tac-Toe
- Pinch and Spoon
- Ten Count
- Match and Take

<u>Measurement and Data</u>
- My Shoes
- Bubbles Bubbles
- Math Pong
- Race to Midnight

<u>Geometry</u>
- Geoboard Shapes
- Tangrams
- Composing Hexagons
- Sigmund Square

COUNTING AND CARDINALITY

Count and Ring the Bell

MATERIALS

210 counters

21 paper plates or trays

Number cards (see page 98)

Bell

HELPFUL HINTS

Using counters that are all the same size and color helps young students focus on counting without the potential distraction of various other attributes. When students are ready, however, using different colors and sizes can offer appropriate challenges. Additionally, the arrangement of the counters offers increasing levels of difficulty (linear, array, circle, scattered).

COMMON CORE STATE STANDARDS IN ACTION:

Math Content: Counting and Cardinality

- Know number names and the count sequence.
- Count to tell the number of objects.
- Compare numbers.

Math Practices

- MP 6 Attend to precision.
- MP 7 Look for and make use of structure.

Count and Ring the Bell

DIRECTIONS

1. Shuffle the number cards and place face down in a pile. Each player takes a plate.

2. The first player flips over a number card.

3. All players race to place the matching number of counters on their plates.

4. The first player to have the correct number of counters on his plate rings the bell.

5. The winning player leaves his plate with counters on the table. All other players take new plates to continue playing.

6. As the plates are left on the table, players should begin placing these in order from least to greatest.

GET STUDENTS TALKING ABOUT MATH

- How did you arrange your counters to help you keep track?
- Which plate has more counters?
- How does this plate of counters relate to your other plate of counters?
- Is there a more efficient way to count the objects?

★ CHALLENGE

Try arranging the counters in a circle prior to counting or try a scattered arrangement.

COUNTING AND CARDINALITY

Stop and Go Counting

MATERIALS

Stop and Go signs (see page 99)

Popsicle sticks/paint stir sticks

Tape

HELPFUL HINTS

Use red paper for the Stop sign.

Use green paper for the Go sign.

Tape the signs (back to back) with the Popsicle stick between the signs.

COMMON CORE STATE STANDARDS IN ACTION

Math Content: Counting and Cardinality

- Know number names and the count sequence.

Math Practices

- MP 3 Construct viable arguments and critique the reasoning of others.
- MP 7 Look for and make use of structure.

Stop and Go Counting

DIRECTIONS

1. One player is the verbal counter while the other player holds the Stop/Go sign.

2. The counter begins counting by ones. As long as the numbers are correct, the other player displays the Go sign.

3. When the counter intentionally makes a mistake in counting, the other player displays the Stop sign. The counter corrects and continues counting.

4. Players switch roles and continue playing.

GET STUDENTS TALKING ABOUT MATH

- Did you hear a mistake? Why is it a mistake?
- Which number should come next?
- What patterns in the counting did you notice?

★ CHALLENGE

Try counting on by tens or counting back by ones.

COUNTING AND CARDINALITY

Is It More or Less?

MATERIALS

Deck of playing cards (without face cards)

HELPFUL HINTS

Use cards that have numbers that are easy to read.

COMMON CORE STATE STANDARDS IN ACTION

Math Content: Counting and Cardinality

- Compare numbers.

Math Practices

- MP 2 Reason abstractly and quantitatively.
- MP 4 Model with mathematics.

Is It More or Less?

DIRECTIONS

1. Each player chooses one card and displays it.
2. Players say the more than, less than, or equal to statement. For example, "Seven is more than five."
3. The player who has the greater number takes both cards.
4. If the cards are equal, each player keeps his own card.
5. Continue playing through the deck. The winner is the player with the most cards.

GET STUDENTS TALKING ABOUT MATH

- Which card is less than the other card?
- How much more is the greater number?
- How can we represent the comparison mathematically?

★ CHALLENGE

Choose two cards, add the numbers together, and compare.

COUNTING AND CARDINALITY

Cereal Chains

MATERIALS

String/yarn

Scissors

Cereal (oat loops and color loops)

Six-sided dice

HELPFUL HINTS

Use child-size scissors or precut the string/yarn for this activity.

COMMON CORE STATE STANDARDS IN ACTION

Math Content: Counting and Cardinality

- Compare numbers.

Math Practices

- MP 2 Reason abstractly and quantitatively.
- MP 7 Look for and make use of structure.

Cereal Chains

DIRECTIONS

1. Each person takes a piece of string.
2. Take turns rolling the six-sided die.
3. If you roll a number that is less than 4, string an oat cereal loop.
4. If you roll a number that is more than 3, string a color cereal loop.
5. After you have played several rounds, tie the string together and wear your cereal chain.
6. Compare and contrast the chains.

GET STUDENTS TALKING ABOUT MATH

- If you roll 5, which cereal should you put on the string?
- Who rolled more numbers that were greater than 3? How do you know?
- What would the chain look like if you rolled 1, 2, 1, 2, 1, 2, 1, 2, 1, 2?

★ CHALLENGE

Try stringing oat cereal loops for even numbers and color cereal loops for odd numbers.

OPERATIONS AND ALGEBRAIC THINKING

Let's Shake It Up

MATERIALS

Beans painted blue on one side (or two-colored counters)

Cans (film canisters, condiment containers, or small cups)

HELPFUL HINTS

Paint large lima beans prior to Family Math Night.

When painting, spread beans out on paper or in a large cardboard box.

Use non-toxic spray paint (blue) to coat only one side of the beans.

Allow painted beans to dry in a well-ventilated area.

COMMON CORE STATE STANDARDS IN ACTION

Math Content: Operations and Algebraic Thinking

- Represent and solve problems involving addition and subtraction.
- Add and subtract within 20.

Math Practices

- MP 1 Make sense of problems and persevere in solving them.
- MP 2 Reason abstractly and quantitatively.

Let's Shake It Up

DIRECTIONS

1. Count the total number of beans in the can.
2. Shake up the can of beans and pour them out.
3. Count the number of beans showing the blue side.
4. Count the number of beans showing the white side.
5. Record the blue + white addition sentence (equation) in your math journal or on the recording sheet (see page 100).
6. Continue shaking, pouring, counting, and recording.

GET STUDENTS TALKING ABOUT MATH

- If you know the total and the number of blue beans, do you know the number of white beans without counting them?
- How many different combinations are there to show the total number of beans?
- Does your equation make sense?

★ CHALLENGE

**Change the total number of beans and play again.
Try this activity using subtraction ideas.**

Frog Hop Addition

MATERIALS

Small plastic frogs

Number lines (0 to 10, 0 to 20, 0 to 30)

Six-sided dice (one per player)

HELPFUL HINTS

If possible, laminate the number lines prior to Family Math Night.

If available, use different colored frogs and/or number lines.

COMMON CORE STATE STANDARDS IN ACTION

Math Content: Operations and Algebraic Thinking

- Represent and solve problems involving addition and subtraction.
- Add and subtract within 20.

Math Practices

- MP 5 Use appropriate tools strategically.
- MP 6 Attend to precision.

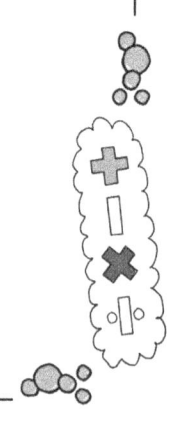

Frog Hop Addition

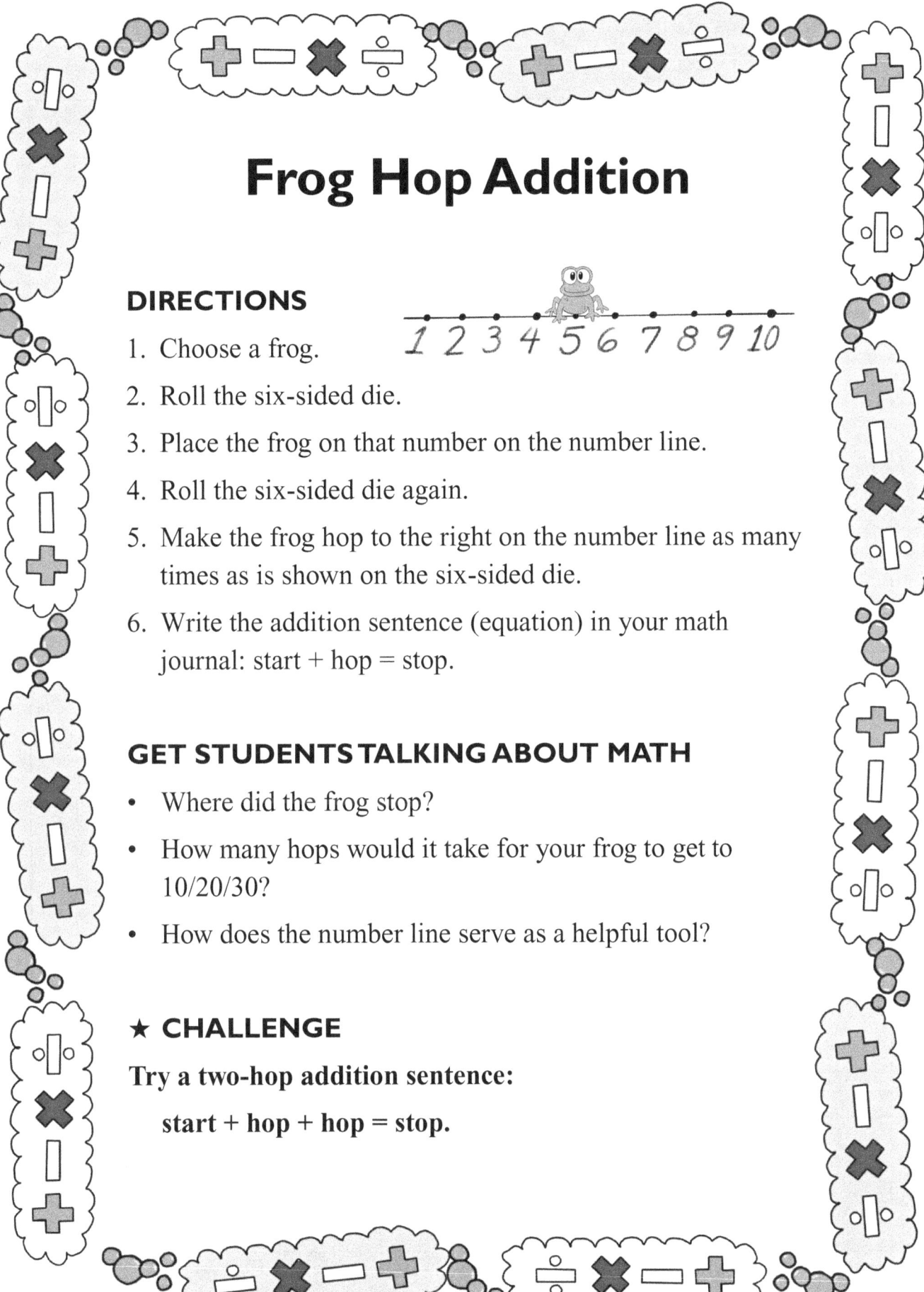

DIRECTIONS

1. Choose a frog.
2. Roll the six-sided die.
3. Place the frog on that number on the number line.
4. Roll the six-sided die again.
5. Make the frog hop to the right on the number line as many times as is shown on the six-sided die.
6. Write the addition sentence (equation) in your math journal: start + hop = stop.

GET STUDENTS TALKING ABOUT MATH

- Where did the frog stop?
- How many hops would it take for your frog to get to 10/20/30?
- How does the number line serve as a helpful tool?

★ CHALLENGE

Try a two-hop addition sentence:

   start + hop + hop = stop.

OPERATIONS AND ALGEBRAIC THINKING

Lu-Lu

MATERIALS

Lu-Lu chips (see page 101)
Hundred chart (see page 102), laminated
Dry-erase markers and erasers

HELPFUL HINTS

Use card stock or plastic counters for the Lu-Lu chips.

Early Hawaiians played Lu-Lu with disks of volcanic stone about 2.5 cm in diameter. These disks served as stone dice called Lu-Lu. The word Lu-Lu means "to shake." The original Lu-Lu chip that represented 1 had the dot placed in the center. Moving the dot to one of the quadrants more clearly represents the numeric relationships. There are four chips in the Lu-Lu set.

COMMON CORE STATE STANDARDS IN ACTION

Math Content: Operations and Algebraic Thinking & Number and Operations in Base Ten

- Represent and solve problems involving addition and subtraction.
- Add and subtract within 20.
- Use place value understanding and properties of operations to add and subtract within 100.

Math Practices

- MP 4 Model with mathematics.
- MP 5 Use appropriate tools strategically.

Lu-Lu

DIRECTIONS

1. The first player shakes all four Lu-Lu chips in both hands and tosses them onto the table.

   Note: A turn consists of two tosses:
   - On the first toss, if all four Lu-Lu chips fall face up (10 dots showing), the player scores 10 points and then tosses all of the Lu-Lu chips again. The number of dots showing on the second toss is added to 10 to get the total score.
   - On the first toss, if one or more of the Lu-Lu chips fall face down, only the face-down Lu-Lu chips are used in the second toss. The score is the total of all the dots showing.

2. The winner is the player with the highest score in a single round, or play can continue (with each player keeping a running total) to an agreed upon score, such as 100.

3. Players can use the hundred charts to keep score.

GET STUDENTS TALKING ABOUT MATH

- How are the dot arrangements related?
- How is the hundred chart helpful?
- What other tools could you use?

★ CHALLENGE

Try playing Lu-Lu with double the amount of chips (two sets).

Magic Squares

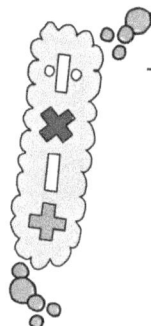

MATERIALS

Sets of dominoes

HELPFUL HINTS

Making a magic square is more challenging using only one set of dominoes because the possibilities are limited.

If available, use sets of dominoes that are different colors to help keep sets organized.

COMMON CORE STATE STANDARDS IN ACTION:

Math Content: Operations and Algebraic Thinking

- Represent and solve problems involving addition and subtraction.
- Understand and apply properties of operations and the relationship between addition and subtraction.
- Add and subtract within 20.
- Work with addition and subtraction equations.

Math Practices

- MP 1 Make sense of problems and persevere in solving them.
- MP 8 Look for and express regularity in repeated reasoning.

Magic Squares

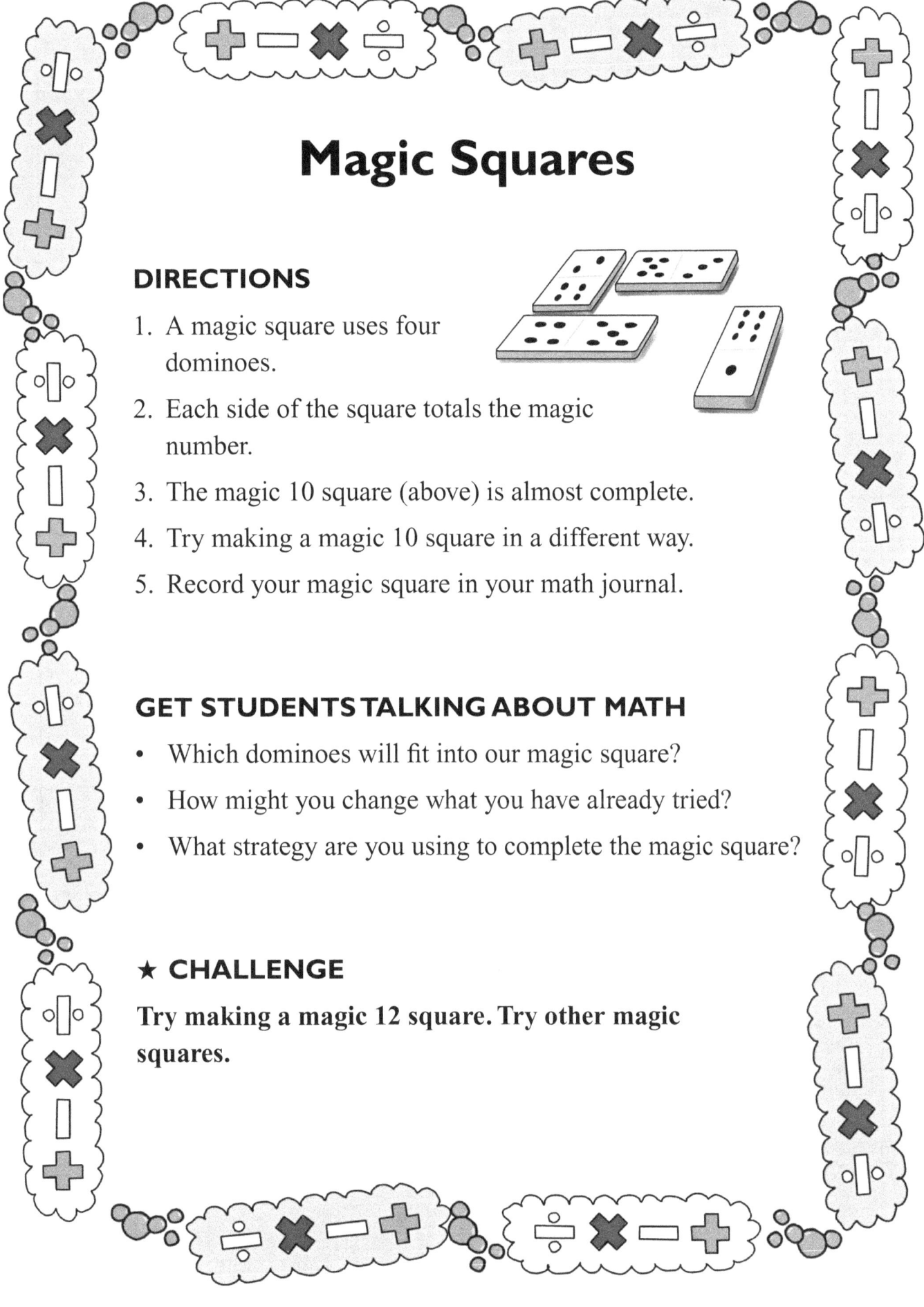

DIRECTIONS

1. A magic square uses four dominoes.
2. Each side of the square totals the magic number.
3. The magic 10 square (above) is almost complete.
4. Try making a magic 10 square in a different way.
5. Record your magic square in your math journal.

GET STUDENTS TALKING ABOUT MATH

- Which dominoes will fit into our magic square?
- How might you change what you have already tried?
- What strategy are you using to complete the magic square?

★ CHALLENGE

Try making a magic 12 square. Try other magic squares.

NUMBER AND OPERATIONS IN BASE TEN

Hundred Chart Tic-Tac-Toe

MATERIALS

Markers or pens (different color for each player)

Paper/dry-erase board/math journal

Hundred chart (see page 102)

HELPFUL HINTS

Encourage students to only use the hundred chart as needed. Flipping the chart over and limiting the number of "peeks" in each game may help students increase ownership of adding and subtracting ones and multiples of 10.

COMMON CORE STATE STANDARDS IN ACTION

Math Content: Number and Operations in Base Ten

- Understand place value.
- Use place value understanding and properties of operations to add and subtract.

Math Practices

- MP 6 Attend to precision.
- MP 8 Look for and express regularity in repeated reasoning.

Note: This activity also appears in *Math Intervention: Building Number Power with Formative Assessments, Differentiation, and Games P–2 and 3–5.*

Hundred Chart Tic-Tac-Toe

DIRECTIONS

1. Each player decides which color marker/pen to use.
2. One player draws a tic-tac-toe board.
3. The first player writes a two-digit number in one of the spaces.
4. The next player uses adding or subtracting ones or tens to write a related number in one of the spaces. The placement of numbers is relative to the position on a hundred chart. (*Note*: If needed, players can check the hundred chart.)
5. Players take turns filling in related numbers.
6. The winner is the first player to have three numbers in a horizontal, vertical, or diagonal row. For example:

54	55	56
64	65	66
74	75	76

GET STUDENTS TALKING ABOUT MATH

- What equation did you use to calculate the number?
- What do you notice about the numbers in each horizontal or vertical row?
- How does your understanding of place value help you know where to place the numbers?

★ CHALLENGE

Try without using the hundred chart or add additional lines and play four in a row.

NUMBER AND OPERATIONS IN BASE TEN

Pinch and Spoon

MATERIALS

Pinch and Spoon game board (see page 103), laminated

Tongs (one set per player)

Spoon (one per player)

Base ten blocks—tens in a bowl, ones in a bowl

Dry-erase markers and erasers

HELPFUL HINTS

Use tongs that can pinch up to 9 "ten" blocks.

Use a spoon that can hold up to 9 "one" blocks.

COMMON CORE STATE STANDARDS IN ACTION

Math Content: Number and Operations in Base Ten

- Understand place value.

Math Practices

- MP 2 Reason abstractly and quantitatively.
- MP 6 Attend to precision.

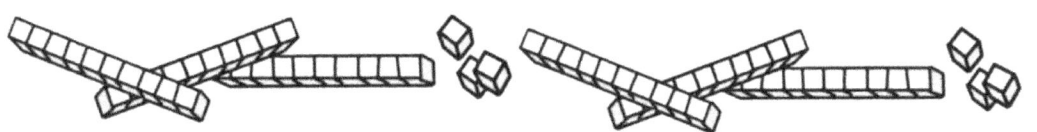

Pinch and Spoon

DIRECTIONS

1. Without looking at the bowls, the first player uses the tongs to "pinch" and remove some "ten" blocks from the bowl. The player also "spoons" some "one" blocks from the bowl.

2. This player determines the value and places his initials on that value on the game board. For example:

    32, or 30 + 2, or 3 tens and 2 ones.

3. Players take turns repeating the process.

4. If a value is pinched and spooned that is not on the game board (or already initialed), the player loses her turn.

5. The winner is the player with four initials in a horizontal or vertical row.

GET STUDENTS TALKING ABOUT MATH

- How is the value represented?
- What is another way to represent the value?
- What if there were two more/less tens?

★ CHALLENGE

Pinch and spoon twice. Add the values.

NUMBER AND OPERATIONS IN BASE TEN

Ten Count

MATERIALS

2 decahedron dice (labeled 1–9)

Base ten blocks (tens and ones)

Ten Count game board (see page 104), laminated

Dry-erase markers (different color for each player) and erasers

HELPFUL HINTS

Using dry-erase markers on a laminated game board is necessary because the activity includes the erasing of numbers.

The game board includes all number lines. Number lines should not be cut apart.

COMMON CORE STATE STANDARDS IN ACTION

Math Content: Number and Operations in Base Ten

- Use place value understanding and properties of operations to add and subtract.

Math Practices

- MP 4 Model with mathematics.
- MP 5 Use appropriate tools strategically.

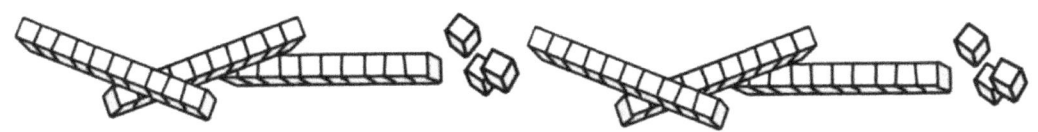

Ten Count

DIRECTIONS

1. The first player rolls two decahedron dice to reveal the tens digit and the ones digit.
2. The player represents this two-digit number with base ten blocks.
3. The player finds where his number belongs on any number line on the game board. For example, he writes 77 on the second number line two spaces after 57.
4. If needed, the player can model skip counting by tens with the base ten blocks.
5. The player writes the number on the number line.
6. If the number rolled does not belong on any number line, the player loses his turn. If the number was previously written by the other player, the player "steals" the spot by erasing and rewriting the number with his marker.
7. Players take turns repeating the process.
8. Ten points are earned for each completed number line on which a player has the most numbers.

GET STUDENTS TALKING ABOUT MATH

- Which equations match the values on the number line?
- How can this two-digit number be represented?
- How are base ten blocks helpful?

★ CHALLENGE

Make your own number lines and try skip counting by 20.

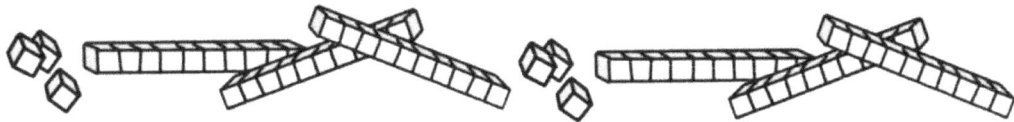

Primary Stations 31

NUMBER AND OPERATIONS IN BASE TEN

Match and Take

MATERIALS

Expression cards (see page 106)

Number lines (see page 105), laminated

Dry-erase markers and erasers

HELPFUL HINTS

Copy the expression cards on card stock.

Laminate number lines.

This game is similar to the game of "Concentration."

COMMON CORE STATE STANDARDS IN ACTION

Math Content: Number and Operations in Base Ten

- Use place value understanding and properties of operations to add and subtract.

Math Practices

- MP 1 Make sense of problems and persevere in solving them.
- MP 3 Construct viable arguments and critique the reasoning of others.

Match and Take

DIRECTIONS

1. Place all expression cards face down in rows.
2. The first player flips over a ▷ card. The player solves the expression by mentally calculating the tens and ones. For example, $34 + 25 = 50 + 9$ because $30 + 20 = 50$ and $4 + 5 = 9$.
3. The player flips over two ▷ cards to try to find the matching expression.
4. If the player has a match, he keeps both cards and takes a second turn.
5. If there is not a match, the player flips all of the cards back over face down.
6. Players continue taking turns until all the cards are taken.
7. Players can use the number line to help with subtraction. For example, $57 - 28 = 30 - 1$ because $50 - 20 = 30$ and $7 - 8 = -1$.
8. The winner is the player with the most cards.

GET STUDENTS TALKING ABOUT MATH

- What is the relationship between the expressions?
- How can you prove the expressions are equivalent?
- What question would you ask if someone incorrectly matched expressions?

★ CHALLENGE

Try making your own equivalent expression cards.

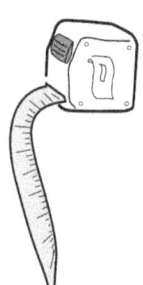

My Shoes

MATERIALS

Connecting links (sometimes called chain links)

HELPFUL HINTS

Avoid placing too many links at this station.

Participants only need to measure the length of their shoes or arm lengths (challenge).

COMMON CORE STATE STANDARDS IN ACTION

Math Content: Measurement and Data

- Measure lengths indirectly and by iterating length units.

Math Practices

- MP 6 Attend to precision.
- MP 7 Look for and make use of structure.

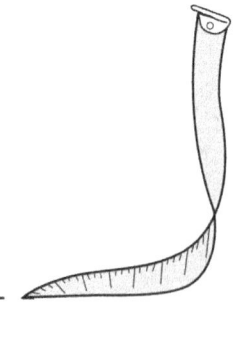

My Shoes

DIRECTIONS

1. Take off your shoes.

2. Measure the length of your shoe using connecting links.

3. Remember, length involves one side, not the entire perimeter (around the shoe).

4. Do you think that your other shoe is the same length? Measure to verify.

5. Record the length of your shoe(s) in your math journal. For example, "My shoe is eight links long."

6. Compare and order several different shoe lengths.

GET STUDENTS TALKING ABOUT MATH

- How many links long is your shoe?
- Are there any patterns in the lengths of our shoes?
- How many links longer/shorter is my shoe than your shoe?

★ CHALLENGE

Measure the length of your arm using links. Compare this length to the length of your shoe.

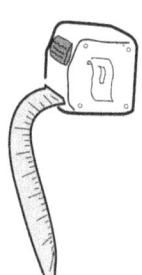

Bubbles Bubbles

MATERIALS

Bubbles and bubble blowers

Paper towels

Trash bins

HELPFUL HINTS

Use small bottles of bubbles. Refill bottles as needed.

Place trash bins nearby.

COMMON CORE STATE STANDARDS IN ACTION

Math Content: Measurement and Data

- Record and interpret data.

Math Practices

- MP 1 Make sense of problems and persevere in solving them.
- MP 2 Reason abstractly and quantitatively.

Bubbles Bubbles

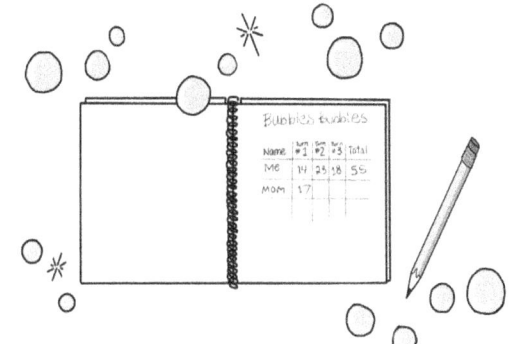

DIRECTIONS

1. Make a table in your math journal.

Example:

Name	Turn #1	Turn #2	Turn #3	Total
John	14	23	18	55
Mom	17			

2. Each person will blow bubbles three times.
3. Record the data and find the totals.

GET STUDENTS TALKING ABOUT MATH

- How many more/fewer bubbles did you blow on Turn #2 than you did on Turn #1?
- Who has the greatest total?
- Looking at the data collected so far, how many bubbles do you think you will blow next?

★ CHALLENGE
Try estimating the total number of bubbles in four turns.

MEASUREMENT AND DATA

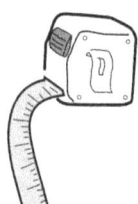

Math Pong

MATERIALS

Math Pong bar graph template (see page 107), laminated

10 plastic cups (per 2 players)

3 bottles of water (per 2 players)

1 ping-pong ball (per 2 players)

Paper towels

2 rulers (inches) and 2 yardsticks

Tape

Dry-erase markers and erasers

HELPFUL HINTS

Have multiple copies of the bar graph template available.

COMMON CORE STATE STANDARDS IN ACTION

Math Content: Measurement and Data

- Measure and estimate lengths in standard units.
- Represent and interpret data.

Math Practices

- MP 4 Model with mathematics.
- MP 8 Look for and express regularity in repeated reasoning.

Math Pong

DIRECTIONS

1. Arrange the cups as shown.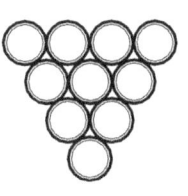
2. Fill each cup with 2 inches of water.
3. Place a line of tape 6 feet away from the cups.
4. The first player throws a ping-pong ball from behind the tape line into a cup. Straight shots and bounce shots are acceptable.
5. The second player records the outcome on the bar graph by shading in one rectangle in the appropriate column: IN, RIM & IN, OUT, or RIM & OUT.
6. Players take turns throwing the ping-pong ball and recording the data. Players discuss how the data change throughout the game.
7. When the ping-pong ball makes it "IN" or "RIM & IN," the player scores 100 points and that cup is removed from the game. The player then gets another turn.
8. Play continues until all the cups are removed. The winner is the player with the highest score.

GET STUDENTS TALKING ABOUT MATH

- How does the bar graph help you organize and analyze the data?
- How many more "OUT" shots than "IN" shots do we have so far?
- What if there were twice as many "RIM & OUT" shots?

★ CHALLENGE

Create a table with data that shows twice as many "IN" as "RIM & OUT" and a total of 28 shots.

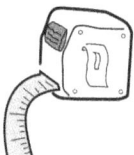

Race to Midnight

MATERIALS

Geared clock for each player

Time Gone By cards (see page 108)

Race to Midnight recording sheet (see page 109), laminated

2 icosahedron dice (20-sided dice for challenge)

Dry-erase markers (fine point) and erasers

HELPFUL HINTS

Have multiple copies of the Race to Midnight recording sheet available. Copy Time Gone By cards on card stock and cut apart.

COMMON CORE STATE STANDARDS IN ACTION

Math Content: Measurement and Data

- Tell and write time.
- Work with time and money.

Math Practices

- MP 3 Construct viable arguments and critique the reasoning of others.
- MP 5 Use appropriate tools strategically.

Race to Midnight

DIRECTIONS

1. Each player sets a clock to 12:00 p.m. (noon).

2. Time Gone By cards are shuffled and placed face down in a pile.

3. The first player draws a card and records the start time, the time gone by (elapsed time), and the end time for that turn. To help find the end time, the player moves the minute hand on her clock. The end time will then serve as the start time for her next turn.

4. Players take turns drawing Time Gone By cards, recording information, setting clocks, and checking one another's work.

5. The winner is the first player to reach midnight or later.

GET STUDENTS TALKING ABOUT MATH

- Does the total number of minutes equal an hour or more?
- Do you agree with the other player's end time? Why or why not?
- How does the clock help you solve the elapsed time problems?

★ CHALLENGE

Roll two icosahedron dice to add seconds to each Time Gone By card drawn.

Geoboard Shapes

MATERIALS

Geoboard

Geoboard geobands (colored rubber bands)

HELPFUL HINTS

Display several geoboards to allow students and parents to compare shapes.

COMMON CORE STATE STANDARDS IN ACTION

Math Content: Geometry

- Identify and describe shapes.
- Analyze, compare, create, and compose shapes.
- Reason with shapes and their attributes.

Math Practices

- MP 3 Construct viable argument and critique the reasoning of others.
- MP 7 Look for and make use of structure.

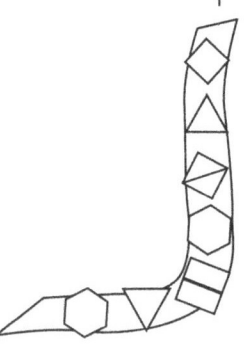

Geoboard Shapes

DIRECTIONS:

1. Use the geobands to make shapes on the geoboard.
2. Try making shapes of different sizes.
3. Try making rectangles (including squares) of different sizes.
4. Try making triangles, hexagons, and pentagons of different sizes.
5. Record some of your shapes in your math journal.

GET STUDENTS TALKING ABOUT MATH

- How many pins does your rectangle touch?
- How many different kinds of triangles can you make?
- Can you make a circle? Why or why not?

★ CHALLENGE

Try seeing how many smaller squares can fit inside a larger square.

Primary Stations

Tangrams

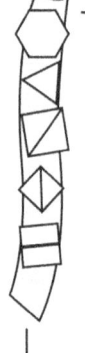

MATERIALS

Sets of tangrams

Answer key for square, trapezoid, butterfly, and whale (see page 110)

Scissors

Colored paper

Pencils

HELPFUL HINTS

Use sets of tangrams that are different colors to help keep sets organized.

Place answer key in a folder or envelope labeled "Answers" so participants can try the tasks before looking at the answers.

COMMON CORE STATE STANDARDS IN ACTION

Math Content: Geometry

- Identify and describe shapes.
- Analyze, compare, create, and compose shapes.
- Reason with shapes and their attributes.

Math Practices

- MP 1 Make sense of problems and persevere in solving them.
- MP 7 Look for and make use of structure.

Tangrams

DIRECTIONS

1. Explore with a set of tangrams. There are seven pieces in a set of tangrams (two large triangles, two small triangles, one medium triangle, one square, and one non-square parallelogram).

2. Compare, contrast, and describe the shapes in the tangram set. Try making a large square using all seven pieces.

3. Try making other shapes, such as a large trapezoid. Try making figures such as a butterfly or a whale. These shapes and figures are in the "Answers" folder/envelope.

4. Record the information in your math journal.

GET STUDENTS TALKING ABOUT MATH

- How many ways can you make a rectangle using all or some of the tangram pieces?
- What other shapes can you make?
- What relationships do you notice between the shapes?

★ CHALLENGE

Trace and cut out the pieces to make your own set of tangrams.

Composing Hexagons

MATERIALS

Composing Hexagons game board (see page 111 or 112)

Pattern blocks name chart (see page 113)

Pattern blocks (original set with or without the extended set, depending on the game board used)

Paper clip

HELPFUL HINTS

If using the original pattern block set, use the game board found on page 111. If using the original set with the extended pattern block set, use the game board found on page 112.

Color the shapes on the Pattern Blocks Name chart and the spinners to match the pattern blocks or download colour copies at www.sigmundsquare.com.

COMMON CORE STATE STANDARDS IN ACTION

Math Content: Geometry

- Identify and describe shapes.
- Analyze, compare, create, and compose shapes.
- Reason with shapes and their attributes.

Math Practices

- MP 6 Attend to precision.
- MP 8 Look for and express regularity in repeated reasoning.

Composing Hexagons

DIRECTIONS

1. The first player uses a paper clip and pencil to spin the spinner (located at the center of the game board).
2. When the paper clip stops on a shape on the spinner, the player takes that shape from the pile of pattern blocks.
3. The player can place the shape anywhere on his side of the game board.
4. The next player takes a turn repeating the same process.
5. Players use the chart to name shapes as they compose each hexagon.
6. The winner is the first player to reach the finish with exactly six completely composed hexagons.

GET STUDENTS TALKING ABOUT MATH

- Which math words best describe the shape and its attributes?
- How could you compose two hexagons with exactly five spins?
- How many more spins do you think you need to compose all six hexagons?

★ CHALLENGE

Try composing each of the six hexagons in a different way.

Sigmund Square

MATERIALS

Sigmund Square Finds His Family by Jennifer Taylor-Cox (traditional paperback or e-book)

Sigmund Square question cards (see page 114 or page 115)

Sigmund Square game board (see page 116) (full-color version of the game board can be downloaded at www.sigmundsquare.com)

Different color pawn for each player

HELPFUL HINTS

The e-book version of Sigmund Square Finds His Family can be downloaded at www.sigmundsquare.com and used on a laptop, desktop, Chromebook, Promethean, or Smart Board. Note the interactive features.

Copy the P–2 cards on colored card stock and the 3–5 cards on a different color of card stock.

COMMON CORE STATE STANDARDS IN ACTION

Math Content: Geometry

- Identify and describe shapes.
- Analyze, compare, create, and compose shapes.
- Reason with shapes and their attributes.
- Draw and identify lines and angles, and classify shapes by properties of their lines and angles.
- Classify two-dimensional figures into categories based on their properties.

Math Practices

- MP 1 Make sense of problems and persevere in solving them.
- MP 5 Use appropriate tools strategically.

Sigmund Square

DIRECTIONS

1. Read all or part of Sigmund Square Finds His Family by Jennifer Taylor-Cox.

2. Shuffle the Sigmund Square question/answer cards and place face up in a pile.

3. Place pawns on START. One player draws a card and asks the other player the geometry question. If the player answers correctly, he moves the number of spaces indicated on the card.

4. Players take turns drawing, asking, and answering the geometry questions.

5. The winner is the first player to reach the FINISH square.

GET STUDENTS TALKING ABOUT MATH

- Is there a relationship between each shape and the number of fingers the character has on each hand?
- What tool could you use to identify right angles?
- Can a trapezoid have a right angle? Explain your thinking.

★ CHALLENGE

Try creating additional geometry question/answer cards.

Primary Stations

Chapter 3
Intermediate Stations
Grade 3 Through Grade 5

Operations and Algebraic Thinking
- Snack Shop
- Multiply on the Fly
- Flip Top Division
- Hooray Array

Number and Operations in Base Ten
- Place the Digit
- Card Math
- Palindromes
- Heads Up Decimals

Number and Operations—Fractions
- Where's $\frac{1}{2}$?
- Clothesline Fractions
- Fishing for Fractions
- Bear Slide

Measurement and Data
- Caterpillar Collection
- Honeybees
- Layers of Cake
- Angle Face

Geometry
- Draw What I Say
- Pentominoes
- Three Smiles
- Geometry Page by Page

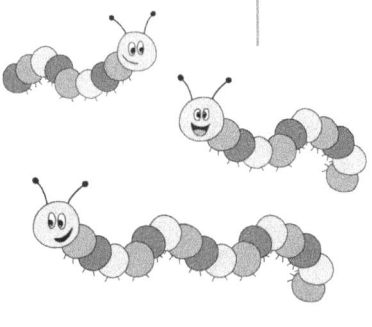

OPERATIONS AND ALGEBRAIC THINKING

Snack Shop

MATERIALS

Snack Shop cards (see page 117), laminated

Dry-erase markers (fine point) and erasers

2 decahedron dice (different colors)

Cuisenaire rods (at least 12 of each color rod)

HELPFUL HINTS

To help students solve multiplicative comparison problems, encourage the use of manipulatives, drawings, and equations.

If Cuisenaire rods are not available, paper Cuisenaire rods can be cut from card stock (lengths: white is 1 cm, red is 2 cm, lime green is 3 cm, purple is 4 cm, yellow is 5 cm, dark green is 6 cm, black is 7 cm, brown is 8 cm, blue is 9 cm, and orange is 10 cm).

COMMON CORE STATE STANDARDS IN ACTION

Math Content: Operations and Algebraic Thinking

- Represent and solve problems involving multiplication and division.
- Multiply and divide within 100.
- Use the four operations with whole numbers to solve problems.

Math Practice

- MP 1 Make sense of problems and persevere in solving them.
- MP 2 Reason abstractly and quantitatively.

Snack Shop

1 (white)
2 (red)
3 (lime green)
4 (purple)
5 (yellow)
6 (dark green)
7 (black)
8 (brown)
9 (blue)
10 (orange)

DIRECTIONS

1. Place Snack Shop cards in a pile face down. Players decide which color dice will represent the cost (in dollars) and which color dice will represent the times more than the amount.

2. The first player draws a card and rolls both dice. He writes and solves the equation to answer the question on the Snack Shop card.

3. While the first player works to solve the problem, the second player models the equation with Cuisenaire rods.

4. Players check the work by comparing the equation to the model. If correct, the first player scores points (in dollars) for the cost of the item. For example, if the hamburger costs five times more than the $2.00 French fries, the player scores $10.00. The card is then placed under the pile.

5. Players switch roles and continue playing. The winner is the player who spends the least (lowest score) after four rounds.

GET STUDENTS TALKING ABOUT MATH

- How does the equation match the model?
- What if the cost for ___ were doubled?
- What if you wanted to buy both items on the card?

★ CHALLENGE

Try using dodecahedron dice (12-sided).

Intermediate Stations

OPERATIONS AND ALGEBRAIC THINKING

Multiply on the Fly

MATERIALS

Six-sided dice

Box lid or deep tray

Dry-erase boards, dry-erase markers and erasers

HELPFUL HINTS

Participants can write equations in math journals or on small dry-erase boards or chalkboards.

The box lid or tray provides a contained area for the rolling of the six-sided dice. If noise is a concern, line the lid or tray with a towel.

COMMON CORE STATE STANDARDS IN ACTION

Math Content: Operations and Algebraic Thinking

- Understand properties of multiplication and the relationship between multiplication and division.
- Multiply and divide within 100.
- Write and interpret numerical expressions.

Math Practices

- MP 1 Make sense of problems and persevere in solving them.
- MP 8 Look for and express regularity in repeated reasoning.

54   *Intermediate Stations*

Multiply on the Fly

DIRECTIONS

1. One person rolls three six-sided dice into the box lid or tray.

2. All players race to write the multiplication equation on a board or in math journals. For example, 3 × 2 × 5 = 30.

3. The first person with an accurate equation receives 10 points.

4. All other players with an accurate equation receive 5 points.

5. Play continues until someone reaches 100 total points.

GET STUDENTS TALKING ABOUT MATH

- What was the multiplication strategy that you used?
- If you change the order of the factors, will you still get the same product?
- What is the highest/lowest product possible each time the three six-sided dice are rolled?

★ CHALLENGE

Try this activity using five six-sided dice.

OPERATIONS AND ALGEBRAIC THINKING

Flip Top Division

MATERIALS

35 plastic bottle caps labeled with dividends (see page 118)

Bowl

Spoon for each player

Calculator

HELPFUL HINTS

Glue dividend circles inside bottle caps or write dividends inside bottle caps. These labels can be inside or outside of the bottle cap.

COMMON CORE STATE STANDARDS IN ACTION

Math Content: Operations and Algebraic Thinking

- Represent and solve problems using multiplication and division.
- Understand properties of multiplication and the relationship between multiplication and division.
- Multiply and divide within 100.
- Use the four operations with whole numbers to solve problems.

Math Practices

- MP 3 Construct viable arguments and critique the reasoning of others.
- MP 4 Model with mathematics.

Flip Top Division

DIRECTIONS

1. Place bottle caps in a bowl.
2. The first player uses a spoon to scoop one bottle cap. The player uses the spoon to "flip" the cap into the air and onto the table to reveal the dividend.
3. If the cap lands revealing the dividend, the player uses the dividend in an accurate division equation. For example, a bottle cap showing a dividend of 24 can be represented as $24 \div 6 = 4$. The player keeps the cap to add to his score.
4. If the cap lands in a way that does not reveal the dividend, the player loses one point. The player continues trying to flip the cap, subtracting one point from his score for each unsuccessful attempt to reveal the dividend.
5. Players take turns flipping caps and stating division equations.
6. Each player keeps a running total of her dividends, minus points for each unsuccessful attempt to reveal the dividend on the bottle cap. The winner is the player with the highest total after 10 rounds. A calculator can be used to check equations.

GET STUDENTS TALKING ABOUT MATH

- Do you agree with the division equation?
- How are division and multiplication related?
- What model could you construct to represent the problem?

★ CHALLENGE

Score double points if the player can create two different division equations. For example, $24 \div 6 = 4$ and $24 \div 12 = 2$.

Intermediate Stations 57

OPERATIONS AND ALGEBRAIC THINKING

Hooray Array

MATERIALS

Connecting cubes (multilinks or snap cubes)

HELPFUL HINTS

Display examples of arrays.

COMMON CORE STATE STANDARDS IN ACTION

Math Content: Operations and Algebraic Thinking

- Represent and solve problems using multiplication and division.
- Understand properties of multiplication and the relationship between multiplication and division.
- Multiply and divide within 100.

Math Practices

- MP 2 Reason abstractly and quantitatively.
- MP 5 Use appropriate tools strategically.

Hooray Array

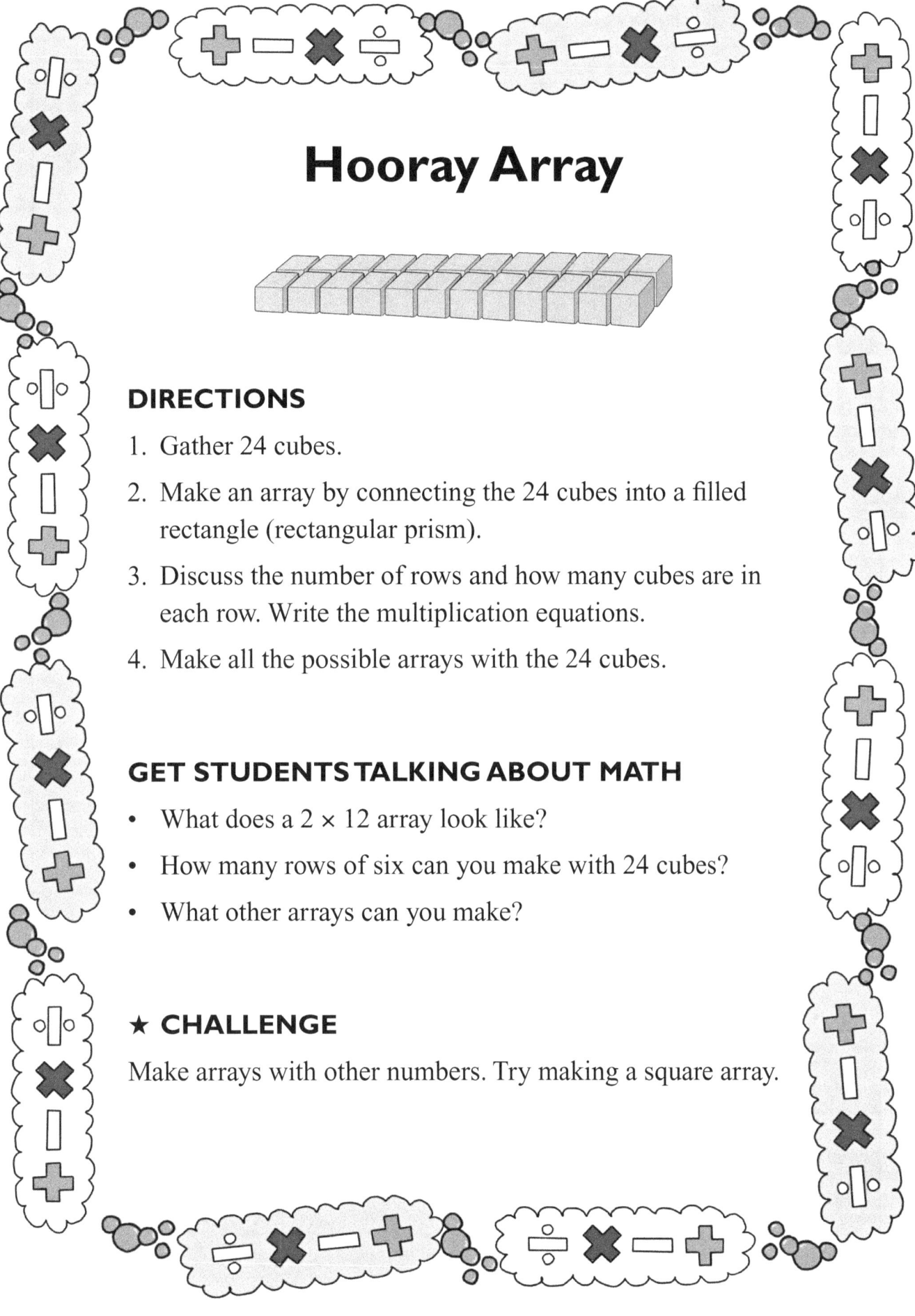

DIRECTIONS

1. Gather 24 cubes.

2. Make an array by connecting the 24 cubes into a filled rectangle (rectangular prism).

3. Discuss the number of rows and how many cubes are in each row. Write the multiplication equations.

4. Make all the possible arrays with the 24 cubes.

GET STUDENTS TALKING ABOUT MATH

- What does a 2 × 12 array look like?
- How many rows of six can you make with 24 cubes?
- What other arrays can you make?

★ CHALLENGE

Make arrays with other numbers. Try making a square array.

NUMBER AND OPERATIONS IN BASE TEN

Place the Digit

MATERIALS

Place the Digit game boards (see page 119), cut out and laminated

Different color dry-erase marker for each player

Erasers

1 decahedron die (per 2 players)

HELPFUL HINTS

There are different versions of this activity (addition, subtraction, and multiplication).

COMMON CORE STATE STANDARDS IN ACTION

Math Content: Number and Operations in Base Ten

- Use place value understanding and properties of operations to perform multi-digit arithmetic.
- Perform operations with multi-digit whole numbers and with decimals to hundredths.

Math Practices

- MP 3 Reason abstractly and quantitatively.
- MP 6 Attend to precision.

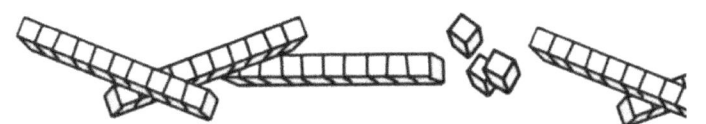 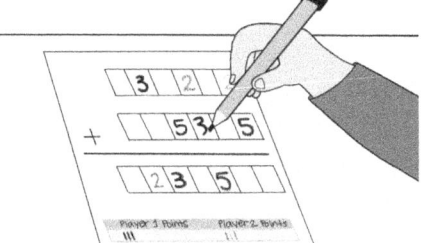

Place the Digit

DIRECTIONS

1. Players share a game board and decahedron die. Each player uses a different colored dry-erase marker.
2. The first player rolls the die and writes the digit in one, two, or three spaces on the game board. One point is earned for each time the digit is written.
3. Players take turns rolling and writing the digit in one, two, or three spaces.
4. Players check for accuracy each time digits are written.
5. As the game progresses, there are fewer options for placing the digits because each equation must be accurate.
6. If the digit rolled can only be placed in two spaces, 2 points are earned. If the digit rolled can only be written in one space, 1 point is earned. If the digit rolled cannot be written in any space, 0 points are earned.
7. The winner is the player with the most points when the game board is complete. (Note: If an error is discovered, the player can erase the digit and corresponding point.)

GET STUDENTS TALKING ABOUT MATH

- Will the digit work in that space? Why or why not?
- Is the equation accurate? Explain your thinking?
- Are there any other digits that could be placed there?

★ CHALLENGE

Try the multiplication version of this game.

 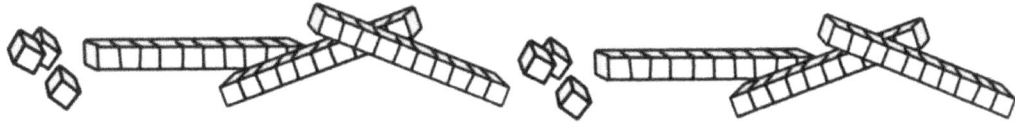

NUMBER AND OPERATIONS IN BASE TEN

Card Math

MATERIALS

Card Math game boards (see page 123), laminated. Enlarge when photocopying so that each card on the board is the size of a playing card.

Deck of cards (jokers removed)

Dry-erase markers and erasers

HELPFUL HINTS

Display an example equation with the cards on the game board.

COMMON CORE STATE STANDARDS IN ACTION

Math Content: Number and Operations in Base Ten

- Use place value understanding and properties of operations to perform multi-digit arithmetic.
- Perform operations with multi-digit whole numbers and with decimals to hundredths.

Math Practices

- MP 1 Make sense of problems and persevere in solving them.
- MP 6 Attend to precision.

Card Math

DIRECTIONS

1. Each player takes eight cards from the pile.
2. Each player arranges the cards on her own game board and writes + or – in the ◯ to make the equation true.
3. Players can rearrange their cards as needed.
4. Face cards and tens are "wild" and can be used for any digit.
5. Players score 100 points for a correct equation. Subtract 10 points for each "wild" card used.
6. If necessary, a player can trade a card for the next card in the deck. Subtract 20 points for each traded card.
7. Players add the scores from each round. The winner is the first to score 1,000 or more.

GET STUDENTS TALKING ABOUT MATH

- What strategy did you use to arrange the cards?
- Which digits will make the equation true?
- How many more points do you need to reach 1,000?

★ CHALLENGE

Try multi-digit addition and subtraction using 12 or more cards without the game board.

NUMBER AND OPERATIONS IN BASE TEN

Palindromes

MATERIALS

Five 6-sided dice

Calculator

HELPFUL HINTS

Display examples of one-step, two-step, and three-step palindromic calculations.

COMMON CORE STATE STANDARDS IN ACTION

Math Content: Number and Operations in Base Ten

- Generalize place value understanding for multi-digit whole numbers.
- Use place value understanding and properties of operations to perform multi-digit arithmetic.
- Understand the place value system.

Math Practices

- MP 1 Make sense of problems and persevere in solving them.
- MP 8 Look for and express regularity in repeated reasoning.

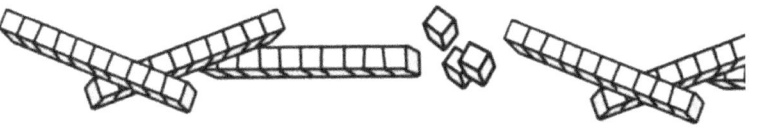

Palindromes

DIRECTIONS

1. A numeric palindrome is a number that reads the same forwards or backwards, such as 535. Some numbers are not palindromes, such as 142. However, if you reverse the digits and then add the numbers, the sum is a palindromic number, 142 + 241 = 383. The number 142 is a one-step palindrome.
2. The first player rolls three dice and arranges the dice into a three-digit number.
3. If the number is a palindrome, the player earns 1 point.
4. If the number is not a palindrome, the player reverses the digits and adds to find the sum, repeating the process up to three steps. A one-step palindrome earns 1 point. A two-step palindrome earns 2 points. A three-step palindrome earns 3 points. If a palindrome cannot be found in three or fewer steps, the player earns 0 points.
5. The other player uses a calculator to check the addition.
6. Players take turns rolling the dice and trying to find palindromic numbers.
7. The winner is the first player to reach 11 points or more.

GET STUDENTS TALKING ABOUT MATH

- What is the relationship between the numbers?
- How many steps do you think it will take to reach a palindromic number?
- What are the generalizations associated with the repeated calculations?

★ CHALLENGE

Try using four or five dice and more steps.

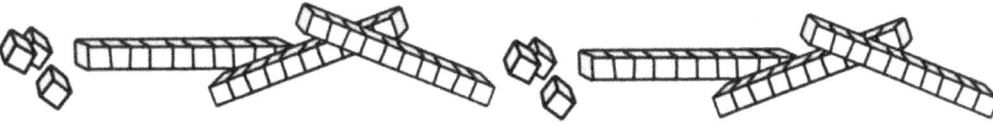

NUMBER AND OPERATIONS IN BASE TEN

Heads Up Decimals

MATERIALS

Heads Up Decimal cards (see page 124)

Decimal grids (see page 125), laminated

Dry-erase markers and erasers

HELPFUL HINTS

Copy decimal cards on card stock (so players cannot see through the paper).

Have several laminated copies of the decimal grids available.

COMMON CORE STATE STANDARDS IN ACTION

Math Content: Number and Operations in Base Ten

- Understand the place value system.
- Perform operations with multi-digit whole numbers and with decimals to hundredths.

Math Content: Number and Operations—Fractions

- Understand decimal notation for fractions and compare decimal fractions.

Math Practices

- MP 4 Model with mathematics.
- MP 7 Look for and make use of structure.

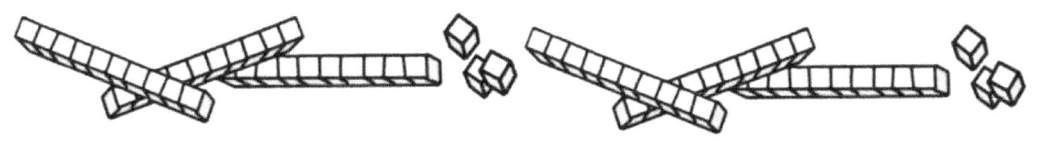

Heads Up Decimals

DIRECTIONS

1. Shuffle and place decimal cards face down in a pile.
2. Each player takes a card and holds it on his forehead so he cannot see the value, but other players can see the decimal fraction value.
3. If a player thinks her value is greater, she says "I am in" and keeps the card held on her forehead.
4. If a player does not think her decimal fraction is greater, she says "I am out" and places her card on the table.
5. Players who are "in" reveal their cards and represent their decimal fraction by shading the decimal grid.
6. Of the "in" players, the person with the greatest decimal fraction earns +1. The "in" players who do not have the greatest decimal fraction earn −1. If "in" players have equivalent decimal fractions, they each earn +0.5. "Out" players earn 0.
7. Play continues with players keeping a running total. The winner is the first player to score 5 or greater.

GET STUDENTS TALKING ABOUT MATH

- How can I represent the value with tenths/hundredths?
- How are tenths and hundredths related?
- What patterns do you see in the values?

★ CHALLENGE

Try using two decimal cards at the same time (add values).

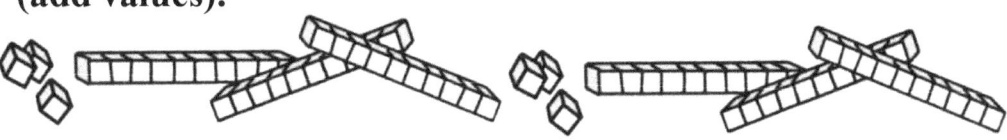

Intermediate Stations

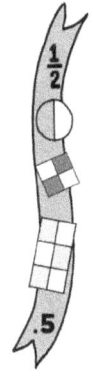

Where's $\frac{1}{2}$?

MATERIALS

Number Line cards (see page 126), laminated

Ruler (inches)

Dry-erase (thick tip) markers (different colors) and erasers

Answers in an envelope

HELPFUL HINTS

Thick markers make darker dots when dropped.

Copy number line cards on card stock.

Cut number line cards apart. Mount on larger paper strips, if needed.

Create answer sheets by copying the number lines and placing $\frac{1}{2}$ in the correct location. Answers can be hidden in an envelope marked "Answers."

COMMON CORE STATE STANDARDS IN ACTION

Math Content: Number and Operations—Fractions

- Develop an understanding of fractions as numbers.
- Extend understanding of fraction equivalence and ordering.

Math Practices

- MP 3 Construct viable arguments and critique the reasoning of others.
- MP 4 Model with mathematics.

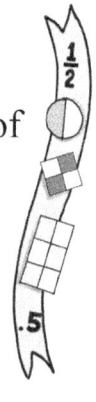

Where's $\frac{1}{2}$?

DIRECTIONS

1. Stack the number line cards face down in a pile.

2. The first player takes a number line from the top of the pile. He places the number line on the table and decides where $\frac{1}{2}$ is located on the number line. He holds his uncapped marker at least 3 inches above the table and drops it on the number line to make a dot representing the location of $\frac{1}{2}$.

3. The other players use different colored markers to drop and mark the location of $\frac{1}{2}$ on the number line.

4. The player whose dot best represents the accurate location of $\frac{1}{2}$ on the number line earns $\frac{1}{2}$ point. (Note: It may be necessary to draw tick marks on the number line prior to dropping the marker or after dropping the marker in order to critique dot locations and evaluate the accurate location of $\frac{1}{2}$.)

5. Players take turns being the first to drop the marker.

6. Each player keeps a running total. The winner is the first player to score 5 points.

GET STUDENTS TALKING ABOUT MATH

- How do you know where $\frac{1}{2}$ is located?
- Does the reasoning make sense?
- What other models could you use to support your answer?

★ CHALLENGE
Try "Where's $\frac{1}{4}$?"

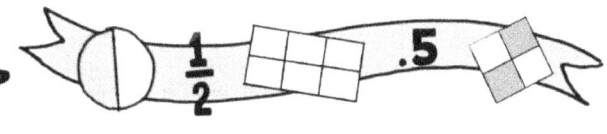

Intermediate Stations 69

Clothesline Fractions

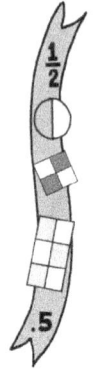

MATERIALS

Index cards

Clothesline (or thick string) and tape, if needed

Clothespins

Markers

Decahedron dice

HELPFUL HINTS

Hang the clothesline in a safe, secure location (against a wall is recommended).

Place a few index card fractions (for example, $\{\frac{1}{3}\}$ and $\{\frac{2}{4}\}$) in the correct order on the clothesline to help start the activity.

COMMON CORE STATE STANDARDS IN ACTION

Math Content: Number and Operations—Fractions

- Develop an understanding of fractions as numbers.
- Extend understanding of fraction equivalence and ordering.

Math Practices

- MP 3 Construct viable arguments and critique the reasoning of others.
- MP 6 Attend to precision.

Clothesline Fractions

DIRECTIONS

1. Roll two decahedron dice. Decide which number will be the numerator and which number will be the denominator.
2. Use a marker to write the fraction on an index card.
3. Use a clothespin to hang the fraction card on the clothesline (which serves as a number line).
4. Take turns rolling decahedron dice, writing fractions, and hanging the cards on the number line.
5. Continue until each person has placed four cards.
6. Make sure the placement and spacing of the fraction cards are accurate and draw the number line in your math journal.

GET STUDENTS TALKING ABOUT MATH

- Do you need to adjust the location of fraction that is already on the number line?
- How do you know where to place the fraction card?
- How could you display equivalent fractions?

★ CHALLENGE

Try using two decahedron dice. Add two numbers to find the numerator and add two numbers to find the denominator.

Fishing for Fractions

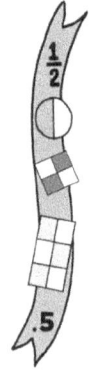

MATERIALS

Fraction fish (see page 129)

Stick and string (to be used as fishing pole)

Magnet (attached at the end of string)

24 paper clips

2 fish baskets (one labeled True and the other labeled False)

HELPFUL HINTS

Cut out fraction fish.

Have blank fish and pencils available for the challenge.

COMMON CORE STATE STANDARDS IN ACTION

Math Content: Number and Operations—Fractions

- Develop an understanding of fractions as numbers.
- Build fractions from unit fractions by applying and extending previous understandings of operations on whole numbers.
- Use equivalent fractions as a strategy to add and subtract fractions.

Math Practices

- MP 2 Reason abstractly and quantitatively.
- MP 7 Look for and make use of structure.

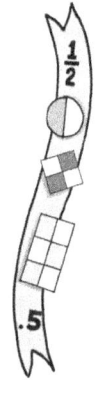

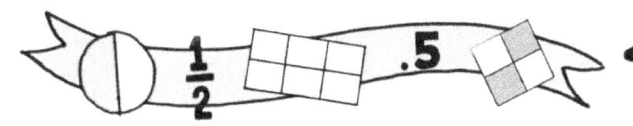

Fishing for Fractions

DIRECTIONS

1. Place a paper clip on each fraction fish. Place fish face down on the table. One player owns the "False" fish basket and the other player owns the "True" fish basket.

2. The first player uses the fishing pole and magnet to "catch" one fish.

3. The player reads the fraction expression or equation and uses reasoning to decide if the fraction fish is True or False. For example, $\frac{1}{2} + \frac{3}{8} = \frac{1}{10}$ is False because $\frac{4}{10} < \frac{1}{2}$.

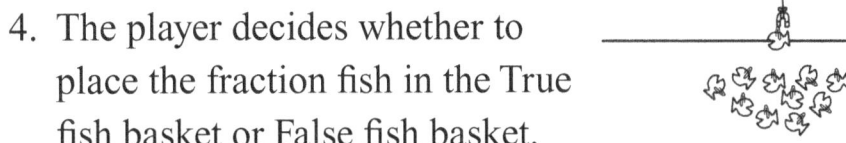

4. The player decides whether to place the fraction fish in the True fish basket or False fish basket.

5. The winner is the player with the most fish in his basket after 10 fish are "caught."

GET STUDENTS TALKING ABOUT MATH

- What is the relationship between the quantities?
- Does the structure of the equation or expression make sense?
- How are the fractions composed?

★ CHALLENGE

Create your own fraction fish. Label True or False.

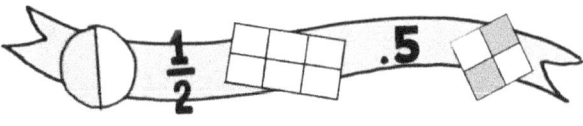

NUMBER AND OPERATIONS—FRACTIONS

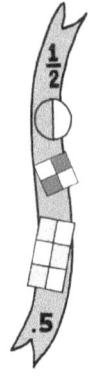

Bear Slide

MATERIALS

4 small bear counters (green, red, blue, yellow)

Bear Slide results cards (see page 130)

Bear Slide number lines (see page 131)

Six-sided die (labeled 1–6)

Dry-erase markers (green, red, blue, yellow) and erasers

HELPFUL HINTS

Copy resources on card stock and cut apart.

Laminate cards and number lines.

COMMON CORE STATE STANDARDS IN ACTION

Math Content: Number and Operations—Fractions

- Build fractions from unit fractions by applying and extending previous understandings of operations on whole numbers.
- Use equivalent fractions as a strategy to add and subtract fractions.
- Apply and extend previous understandings of multiplication and division to multiply and divide fractions.

Math Practices

- MP 5 Use appropriate tools strategically.
- MP 8 Look for and express regularity in repeated reasoning.

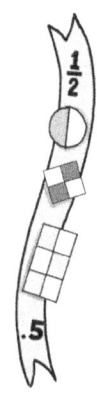

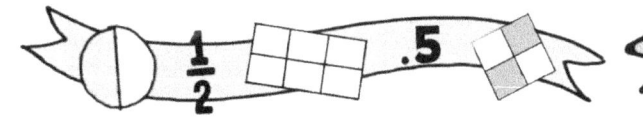

Bear Slide

DIRECTIONS

1. Each player chooses a bear and marker (green, red, blue, or yellow). Place Bear Slide results cards in a pile and flip over one card per round.

2. The bear with the star names the player who rolls the die. The player writes the number rolled (factor) in all four boxes on the card. The player also draws a line on his number line to represent his slide (product).

3. Other players use markers to draw lines on the number lines to represent each bear's slide. The products are determined using "scaling" or "resizing" without performing the indicated multiplication. For example, $\frac{1}{2} \times 4$ = one half of four. The player draws a line representing 2 as the product. As a further example, $1\frac{1}{2} \times 4$ = one four and half of four. The player draws a line representing 6 as the product.

4. The player whose bear has the longest slide wins the round.

5. All players keep a running total of each slide. The winner is the player with the highest total after six rounds.

GET STUDENTS TALKING ABOUT MATH

- How can we use the number line to represent the equation?
- What shortcuts can be used to find the product?
- Are there any generalizations associated with multiplying fractions and whole numbers?

★ CHALLENGE

Try using whole number factors greater than 6.

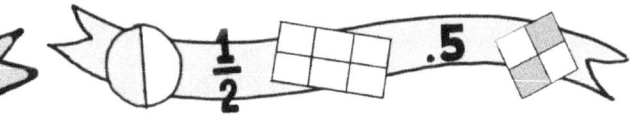

Intermediate Stations 75

MEASUREMENT AND DATA

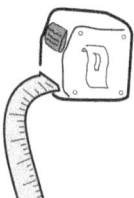

Caterpillar Collection

MATERIALS

Caterpillars (see page 132)

1 pair of tweezers for each player

Stopwatch/timer

Rulers (quarter inch)

HELPFUL HINTS

Copy caterpillars on thick green card stock. Use actual size setting when printing to ensure the caterpillars are the accurate lengths. There are ten caterpillars of each size. There are 12 sizes in quarter inch increments (one quarter inch through three inches).

Cut caterpillars apart.

COMMON CORE STATE STANDARDS IN ACTION

Math Content: Measurement and Data

- Represent and interpret data.

Math Practices

- MP 5 Use appropriate tools strategically.
- MP 6 Attend to precision.

Caterpillar Collection

DIRECTIONS

1. Place pile of caterpillars on the table.
2. Start the timer. Each player uses tweezers to gather a collection of caterpillars. Caterpillars must be gathered one at a time.
3. After one minute, players stop gathering caterpillars and begin sorting the caterpillars by length.
4. Players use rulers to measure their caterpillars to the nearest $\frac{1}{4}$ inch.
5. Each player makes a line plot to display the data set. For example,

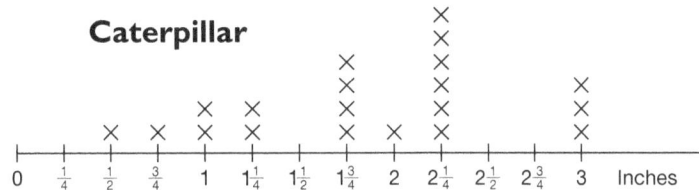

6. Add the lengths of the caterpillars in your collection. Ten points are earned for the greatest total length. Ten points are earned for the most caterpillars. Ten points are earned for an accurate line plot.
7. The winner is the player with the most points.

GET STUDENTS TALKING ABOUT MATH

- How many caterpillars are longer than 2 inches?
- What helps you measure efficiently and accurately?
- What would you estimate is the total of all players' caterpillars' lengths altogether?

★ CHALLENGE

Divide the total length of all the caterpillars in your collection by the number of caterpillars in your collection to determine the mean (one of the measures of central tendency).

MEASUREMENT AND DATA

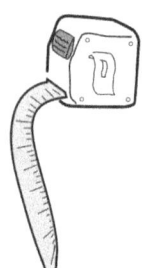

Honeybees

MATERIALS

15 Pattern Block hexagons for each participant

HELPFUL HINTS

Post definitions of area and perimeter:

Area—the measure of space within a figure

Perimeter—the distance around the outside of a figure

COMMON CORE STATE STANDARDS IN ACTION

Math Content: Measurement and Data

- Geometric measurement: understand concepts of area and relate area to multiplication and to addition.

- Geometric measurement: recognize perimeter as an attribute of plane figures and distinguish between linear and area measures.

Math Practices

- MP 2 Reason abstractly and quantitatively.
- MP 8 Look for and express regularity in repeated reasoning.

Honeybees

DIRECTIONS

1. The hives of honeybees are made up of many cells. Each cell is in the shape of a hexagon.

2. Using the pattern blocks, each person constructs a hive that is made up of 15 cells.

3. Find the area (consider one hexagon as one unit) and find the perimeter (consider one side of the hexagon as one unit).

4. Compare the hives. If the areas are equal, will the perimeters also be equal?

5. Try making a 15-cell hive with the largest/smallest perimeter.

GET STUDENTS TALKING ABOUT MATH

- Which hive has the greatest perimeter? Why?
- What repeated reasoning could you use to decrease the perimeter?
- Is it possible to make a symmetrical hive within 15 cells?

★ CHALLENGE

Try making and comparing hives with a larger/smaller number of cells.

Intermediate Stations

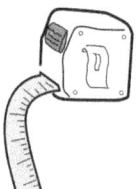

Layers of Cake

MATERIALS

2 decahedron dice

100-centimeter cubes

Centimeter grid paper

Calculator

Pencils

HELPFUL HINTS

Use dry-erase markers if centimeter paper is laminated.

COMMON CORE STATE STANDARDS IN ACTION

Math Content: Measurement and Data

- Geometric measurement: understand concepts of volume and relate volume to multiplication and division.

Math Practices

- MP 4 Model with mathematics.
- MP 7 Look for and make use of structure.

Layers of Cake

DIRECTIONS

1. The first player rolls two decahedron dice and multiplies the numbers. This product serves as the number of centimeters in the bottom layer of the rectangular prism "cake."
2. The player can represent the bottom layer using centimeter cubes, centimeter grid paper, or by writing the number of cubic centimeters.
3. The first player then rolls one decahedron die to learn the number of layers of the rectangular prism "cake" (how many layers).
4. All players calculate the volume of the rectangular prism "cake" (L × W × H, or total bottom layer times height).
5. If the volume is greater than 64 cm^3, the player earns 1 point. If the volume is less than 64 cm^3, the other player earns 1 point. If the volume is exactly 64 cm^3, the player earns 8 points.
6. Players take turns rolling dice and determining layers, heights, and volumes. The winner is the first player to reach 9 points or more.

GET STUDENTS TALKING ABOUT MATH

- How did you represent the quantities?
- Is there another way to represent the volume?
- How is the rectangular prism "cake" composed?

★ CHALLENGE

Try "slices" (horizontal) instead of "layers" (vertical) of a rectangular prism "cake."

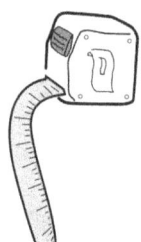

Angle Face

MATERIALS

Angle Face cards (see page 133)

Angle Face sorting mat (see page 134)

Right angle tool

Protractor or angle viewer

HELPFUL HINTS

A right angle tool can be a copy of a thick letter L on a transparency. It can also be a corner of a piece of paper or other representation of 90°.

COMMON CORE STATE STANDARDS IN ACTION

Math Content: Measurement and Data

- Geometric measurement: understand concepts of angle and measure angles.

Math Content: Geometry

- Draw and identify lines and angles, and classify shapes by properties of their lines and angles.

Math Practices

- MP 3 Construct viable arguments and critique the reasoning of others.
- MP 5 Use appropriate tools strategically.

Angle Face

DIRECTIONS

1. Shuffle Angle Face cards and place cards face down in a pile.
2. The first player takes a card and estimates the angle size. The player places the card in the predicted location on his sorting mat. (Note: An acute angle is less than 90°, a right angle is 90°, and an obtuse angle is greater than 90°.)
3. The player uses a tool to measure the angle to the nearest 10°.
4. The player records the measurement as his score.
5. Players take turns taking cards, estimating, placing angles on individual sorting mats, measuring angles to the nearest 10°, and keeping running totals as individual scores.
6. Each time the player draws a card, she decides if she wants to keep the angle card (add to her running total), reject the card (do not add to her running total), or discard one of the cards she drew previously (subtract from her running total). (Note: Rejected and discarded cards are placed underneath the pile of cards.)
7. The winner is the first player to score exactly 360° (total degrees of all angles on the player's sorting mat).

GET STUDENTS TALKING ABOUT MATH

- Do you agree with the other player's estimate of the angle size? Why or why not?
- How does this measurement tool help you?
- How did you measure the angle?

★ CHALLENGE

Try drawing each angle.

GEOMETRY

Draw What I Say

MATERIALS

Draw What I Say cards (see page 135)

Dry-erase boards

Dry-erase markers and erasers

HELPFUL HINTS

Using individual dry-erase boards or clipboards allows the player to draw without others seeing the drawing.

COMMON CORE STATE STANDARDS IN ACTION

Math Content: Geometry

- Reason with shapes and their attributes.
- Draw and identify lines and angles, and classify shapes by properties of their lines and angles.

Math Practices

- MP 3 Construct viable arguments and critique the reasoning of others.
- MP 6 Attend to precision.

Draw What I Say

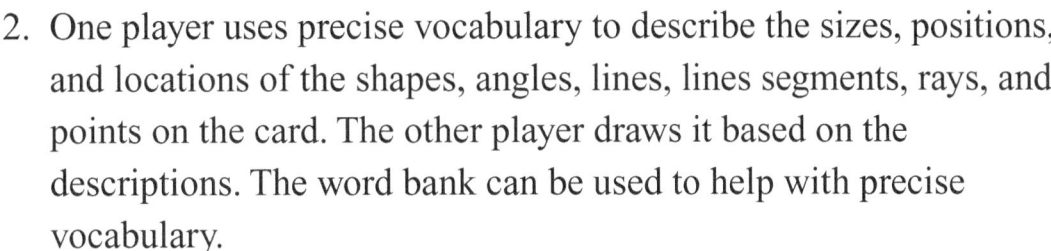

DIRECTIONS

1. Shuffle the Draw What I Say cards and place cards face down in a pile.
2. One player uses precise vocabulary to describe the sizes, positions, and locations of the shapes, angles, lines, lines segments, rays, and points on the card. The other player draws it based on the descriptions. The word bank can be used to help with precise vocabulary.
3. The player drawing cannot look at the card or ask questions. The player describing cannot look at what the other player is drawing.
4. After the drawing is complete, the players compare it to the drawing on the card. The players grade the drawing A, B, C.
5. The winner is the player with the most 'A's after four rounds.

Acute	Angle	Circuit	Hexagon
Horizontal	Line	Line segment	Parallel
Parallelogram	Pentagon	Perpendicular	Point
Obtuse	Octagon	Quadrilateral	Ray
Rectangle	Rhombus	Right	Square
Trapezoid	Triangle	Vertical	

GET STUDENTS TALKING ABOUT MATH

- How can you communicate precisely?
- What mathematical language is most helpful?
- After comparing the drawings, what improvements could be made to the descriptions?

★ CHALLENGE

Allow the player drawing to ask three clarifying questions.

Intermediate Stations 85

GEOMETRY

Pentominoes

MATERIALS

Square tiles

Graph paper

Scissors

Answer key (see page 136)

Folder or envelope

HELPFUL HINTS

Display at least one of the pentominoes as an example.

Post definitions of congruent, area, and perimeter.

Place the answer key in a folder or envelope labeled "Answers" so that participants try constructing several different pentominoes before looking at answers.

COMMON CORE STATE STANDARDS IN ACTION

Math Content: Geometry

- Classify two-dimensional figures into categories based on their properties.

Math Content: Measurement and Data

- Geometric measurement: understand concepts of area and relate area to multiplication and to addition.
- Geometric measurement: recognize perimeter as an attribute of plane figures and distinguish between linear and area measures.

Math Practices

- MP 4 Model with mathematics.
- MP 7 Look for and make use of structure.

Pentominoes

DIRECTIONS

1. Make a pentomino by grouping five squares together so that every square has at least one of its sides in common with at least one side of another square.

2. One example of a pentomino is a rectangle formed by lining up all five squares. Try making more complex pentominoes (such as a "T" form). There are 12 different pentominoes.
   Note: If the pentominoes are congruent (same shape flipped or rotated), they are not considered different.

3. Draw your pentominoes on graph paper and cut them out.

GET STUDENTS TALKING ABOUT MATH

- How do you know that you have made a different pentomino?
- Do all of the pentominoes have the same area? Do all of the pentominoes have the same perimeter?
- Which pentominoes can be folded to make a cube without a lid?

★ CHALLENGE

Try using all 12 pentominoes to construct a 6×10 rectangle.

Three Smiles

MATERIALS

Three Smiles coordinate plane (see page 137)

2 decahedron dice

Dry-erase markers (different colors) and erasers

HELPFUL HINTS

Post reminder of which is the x-axis and which is the y-axis.

COMMON CORE STATE STANDARDS IN ACTION

Math Content: Geometry

- Graph points on the coordinate plane to solve real world and mathematical problems.

Math Practices

- MP 1 Make sense of problems and persevere in solving them.
- MP 7 Look for and make use of structure.

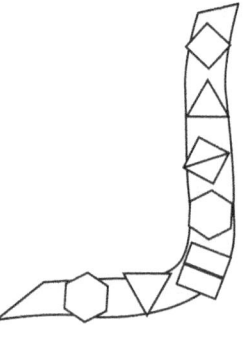

Three Smiles

DIRECTIONS

1. The first player rolls the two decahedron dice and decides which number will serve as the x-coordinate and which number will serve as the y-coordinate.

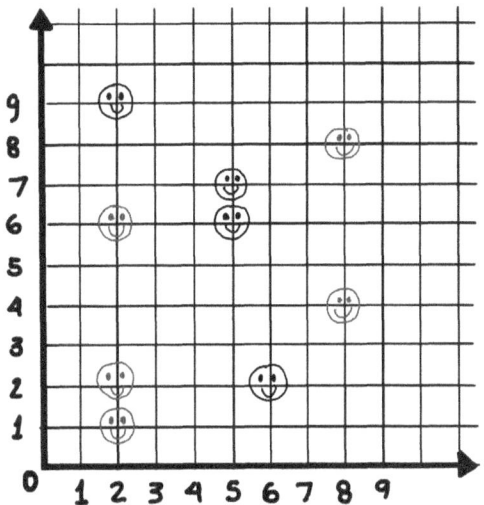

2. The player writes the coordinate pair and uses their marker to draw a smiley face on the intersection of the coordinates on the coordinate plane.

3. The next player rolls the dice and does the same, placing a smiley face on the intersection of the coordinates using their marker.

4. Players take turns graphing points with smiley faces.

5. The winner is the first player with three smiley faces in a row on the coordinate plane (diagonal, vertical, or horizontal).

GET STUDENTS TALKING ABOUT MATH

- What strategy did you use to graph the point?
- What patterns do you notice on the coordinate plane?
- Is there a way to block your opponent?

★ CHALLENGE
Try four smiley faces in a row.

Geometry Page by Page

MATERIALS

Sigmund Square Finds His Family by Jennifer Taylor-Cox (traditional paperback or e-book)

Geometry vocabulary (see page 133)

HELPFUL HINTS

The e-book version of Sigmund Square Finds His Family can be downloaded at www.sigmundsquare.com and used on a laptop, desktop, Chromebook, Promethean, or Smart Board. Note the interactive features.

Cut apart the geometry vocabulary.

COMMON CORE STATE STANDARDS IN ACTION

Math Content: Geometry

- Reason with shapes and their attributes.
- Draw and identify lines and angles, and classify shapes by properties of their lines and angles.
- Classify two-dimensional figures into categories based on their properties.

Math Practices

- MP 4 Model with mathematics.
- MP 6 Attend to precision.

Geometry Page by Page

DIRECTIONS

1. Each player chooses five geometry vocabulary words (with definitions).

2. Watch, or have one person read aloud, *Sigmund Square Finds His Family* by Jennifer Taylor-Cox. Also found at sigmundsquare.com.

3. If a player hears his word or sees a representation of his word on the page, he scores one point. If the player can also say the definition of the word without looking, he scores 2 points.

4. Continue watching/reading the story and stating geometry vocabulary meaning.

5. The winner is the first player to score 10 points.

GET STUDENTS TALKING ABOUT MATH

- How is the math concept represented?
- Which math words best describe the concept?
- How does precise vocabulary help us communicate in math?

★ CHALLENGE

Try choosing more geometry vocabulary words.

Intermediate Stations

Chapter 4
Resources

- Family Math Night Invitation to Parents
- Family Math Night Journal Cover
- Family Math Night Evaluation
- Numeral Cards
- Stop and Go Signs
- Let's Shake It Up
- Lu-Lu Chips
- Hundred Chart
- Pinch and Spoon Game Board
- Ten Count Game Board
- Match and Take Number Lines
- Match and Take Expression Cards
- Math Pong Bar Graph Template
- Race to Midnight Time Gone By Cards
- Race to Midnight Recording Sheet
- Tangrams Answer Key
- Composing Hexagons 1
- Composing Hexagons 2
- Pattern Blocks Name Chart
- Sigmund Square Question Cards P–2
- Sigmund Square Question Cards 3–5
- Sigmund Square Game Board
- Snack Shop Cards
- Flip Top Division Bottle Cap Labels
- Place the Digit Game Boards
- Card Math Game Board
- Heads Up Decimal Cards
- Heads Up Decimal Grid
- Where's $\frac{1}{2}$? Number Lines
- Fraction Fish
- Bear Slide Results Cards
- Bear Slide Number Lines
- Caterpillars
- Angle Face Cards
- Angle Face Sorting Mats
- Draw What I Say Cards
- Pentominoes
- Three Smiles Coordinate Plane
- Geometry Vocabulary

Family Math Night Invitation to Parents

On _____ (date and time),

_____ (school name)

will hold an exciting event called **Family Math Night**!

Students, parents, siblings, and other relatives are invited to attend a fun-filled evening of mathematical pleasure. The intent of **Family Math Night** is to participate in math standards in action as we strengthen the mathematical application, problem solving, and communication skills of students through the power of family interaction.

We encourage you to continue to support your child's mathematical growth through your participation in Family Math Night.

FAMILY MATH NIGHT JOURNAL COVER

Family Math Night at _____
School Name

Date _____

FAMILY MATH NIGHT
Math Standards in Action
MATH JOURNAL

Student's Name _____

FAMILY MATH NIGHT
Math Standards in Action

_____
School Name

Date _____

Did you enjoy the Family Math Night?	
Which activity did you like the best?	
Which activity do you plan to try again at home?	
Would you change anything about Family Math Night?	

Signed (Student) _____

Signed (Parent/Guest) _____

Date _____

NUMBER CARDS

0	1	2	3	4	5	6
7	8	9	10	11	12	13
14	15	16	17	18	19	20

- -

0	1	2	3	4	5	6
7	8	9	10	11	12	13
14	15	16	17	18	19	20

STOP AND GO SIGNS

LET'S SHAKE IT UP

Shake the can and drop the beans. Count the blue beans. Count the white beans. Record the information. Use the same number of beans in the can and repeat.

Blue Beans	White Beans	Total Beans

LU-LU CHIPS

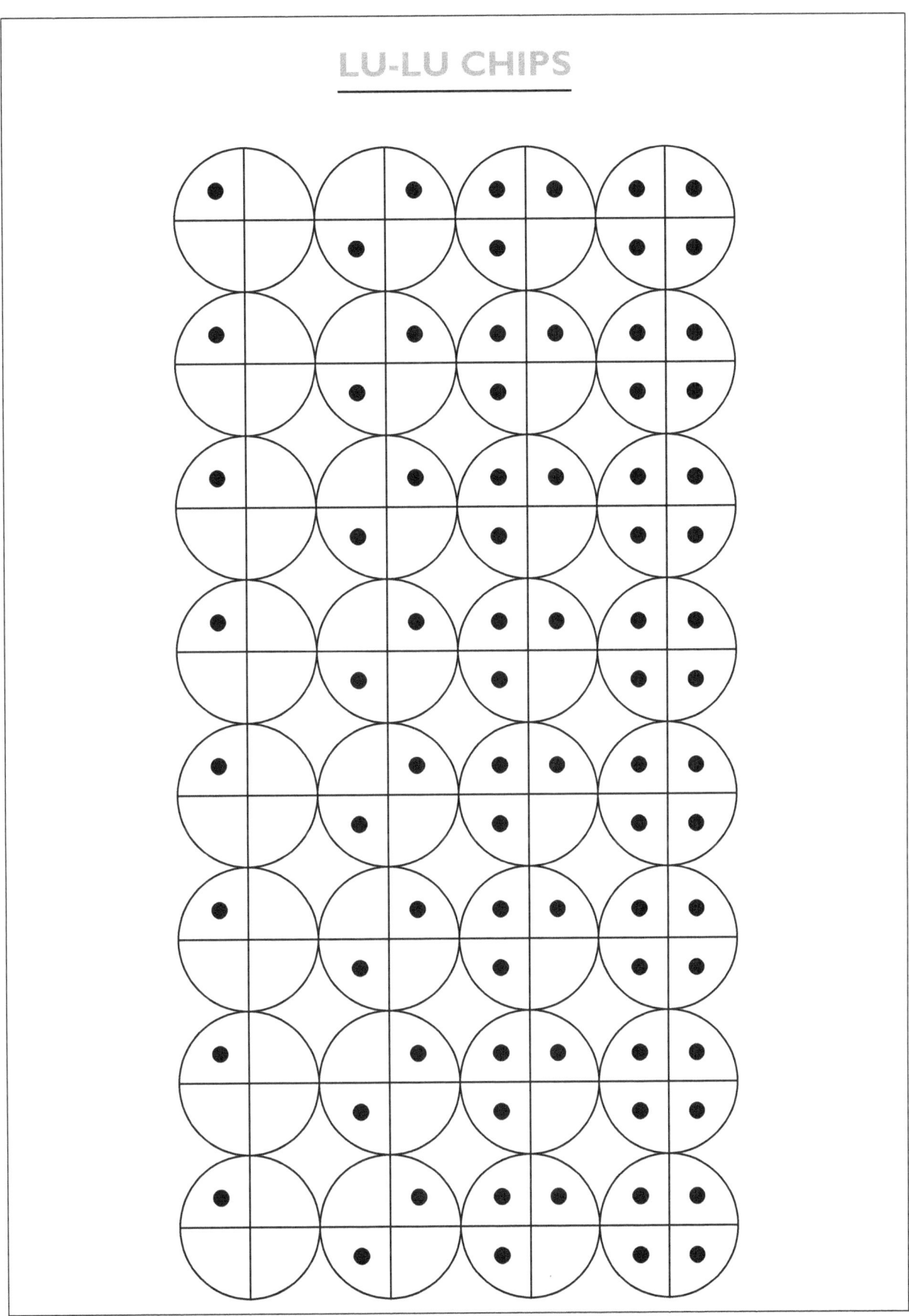

HUNDRED CHART

1	2	3	4	5	6	7	8	9	10
11	12	13	14	15	16	17	18	19	20
21	22	23	24	25	26	27	28	29	30
31	32	33	34	35	36	37	38	39	40
41	42	43	44	45	46	47	48	49	50
51	52	53	54	55	56	57	58	59	60
61	62	63	64	65	66	67	68	69	70
71	72	73	74	75	76	77	78	79	80
81	82	83	84	85	86	87	88	89	90
91	92	93	94	95	96	97	98	99	100

- -

1	2	3	4	5	6	7	8	9	10
11	12	13	14	15	16	17	18	19	20
21	22	23	24	25	26	27	28	29	30
31	32	33	34	35	36	37	38	39	40
41	42	43	44	45	46	47	48	49	50
51	52	53	54	55	56	57	58	59	60
61	62	63	64	65	66	67	68	69	70
71	72	73	74	75	76	77	78	79	80
81	82	83	84	85	86	87	88	89	90
91	92	93	94	95	96	97	98	99	100

PINCH AND SPOON GAME BOARD

12	20 + 3	3 tens 4 ones	45	50 + 6	6 tens 7 ones	78
80 + 9	9 tens 1 one	15	20 + 6	3 tens 7 ones	48	50 + 9
6 tens 9 ones	71	60 + 9	9 tens 3 ones	14	20 + 5	3 tens 6 ones
47	50 + 8	8 tens 2 ones	74	70 + 9	1 ten 3 ones	94
20 + 4	3 tens 5 ones	46	50 + 7	6 tens 8 ones	83	40 + 9
7 tens 2 ones	16	20 + 7	3 tens 8 ones	86	50 + 5	6 tens 1 one
95	80 + 5	9 tens 6 ones	17	20 + 8	3 tens 9 ones	40
50 + 1	6 tens 2 ones	84	70 + 3	8 tens 1 one	18	20 + 9

TEN COUNT GAME BOARD

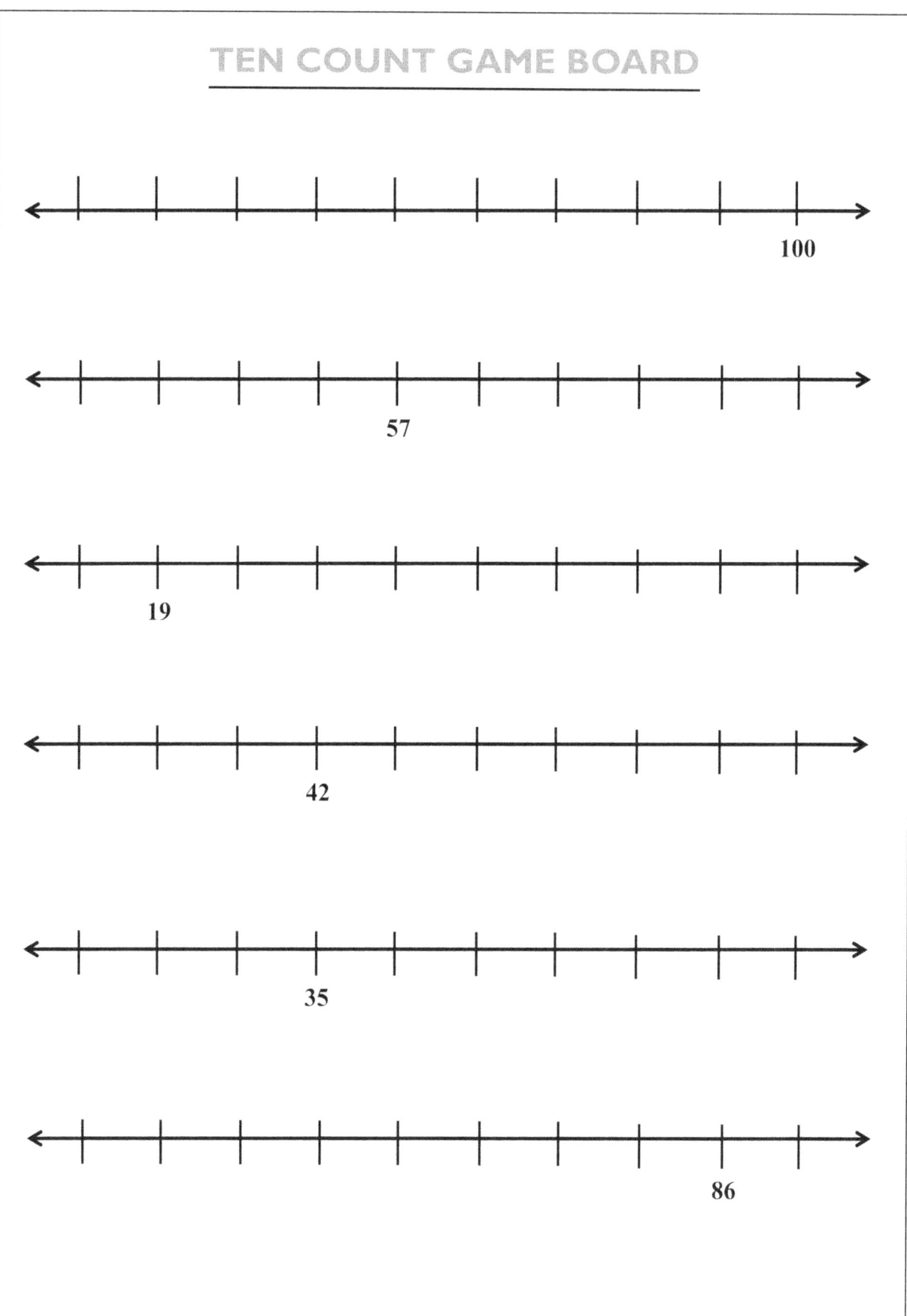

104 © 2017, *Family Math Night*, Jennifer Taylor-Cox, Routledge

MATCH AND TAKE NUMBER LINES

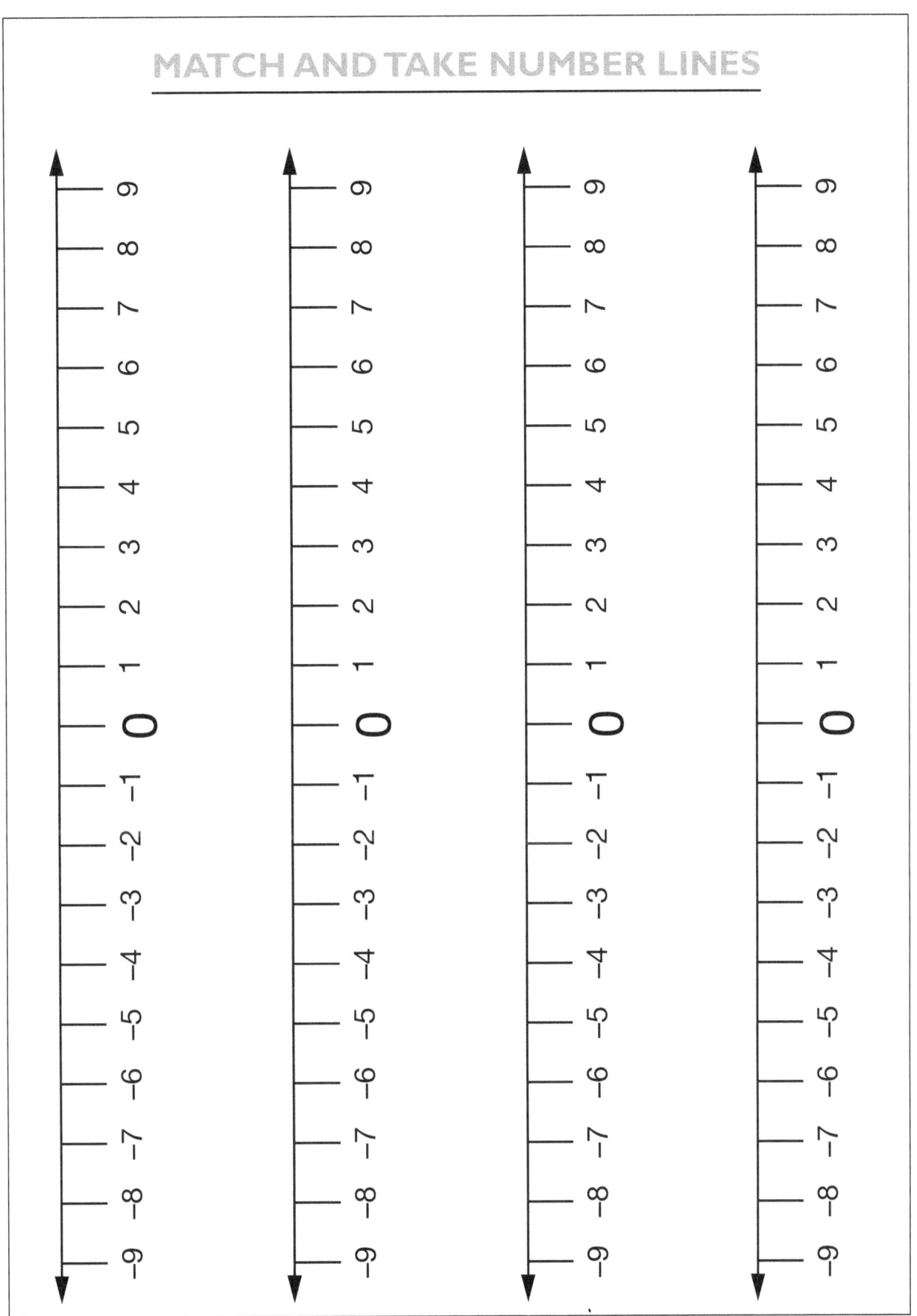

MATCH AND TAKE EXPRESSION CARDS

34 + 25	50 + 9	62 + 14	70 + 6
53 + 31	80 + 4	85 + 2	80 + 7
46 + 25	0 + 11	29 + 24	40 + 13
72 + 18	80 + 10	37 + 27	50 + 14
35 − 12	20 + 3	87 − 53	30 + 4
98 − 67	30 + 1	67 − 25	40 + 2
57 − 28	30 − 1	73 − 25	50 − 2
42 − 36	10 − 4	81 − 46	40 − 5

MATH PONG BAR GRAPH TEMPLATE

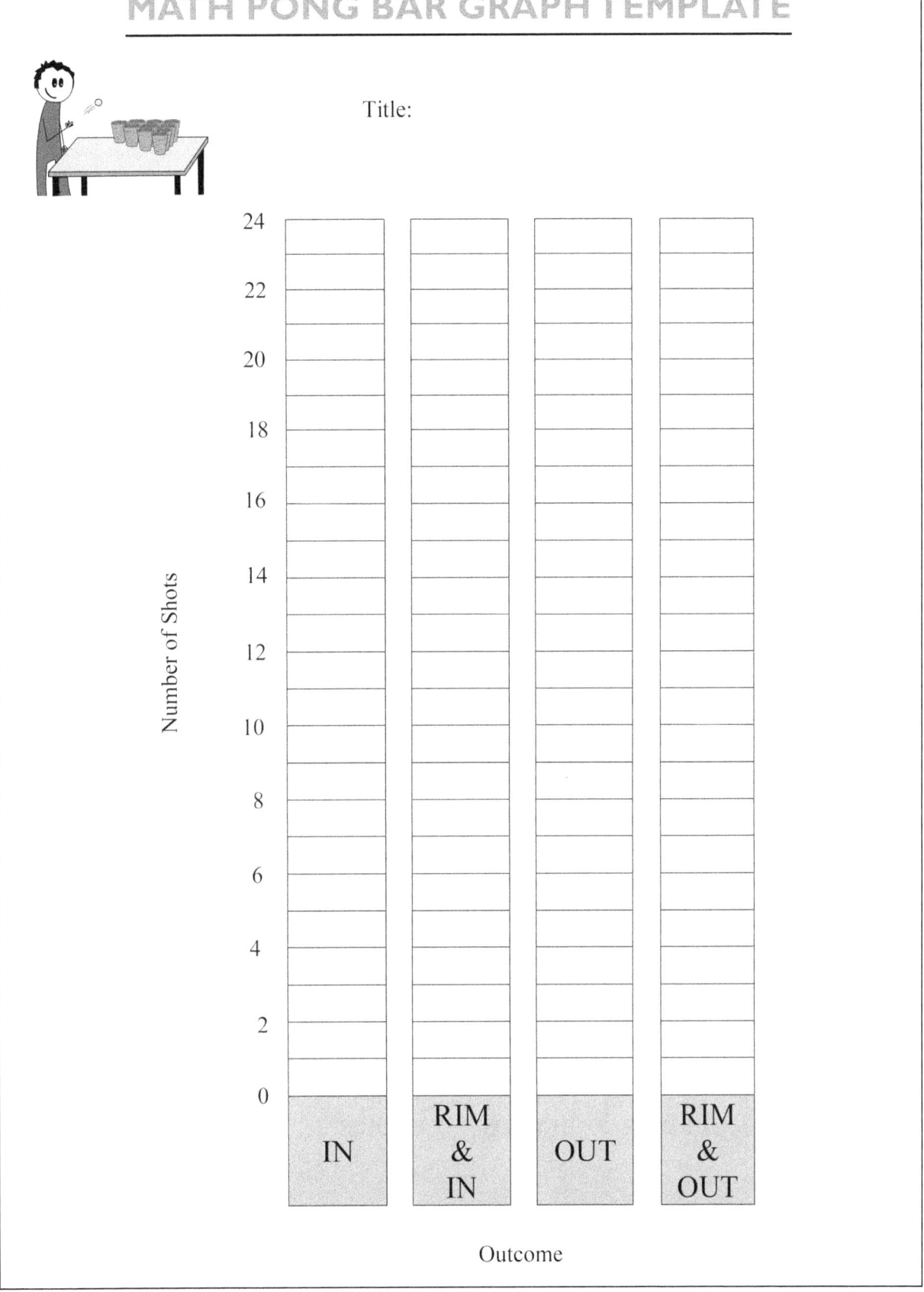

RACE TO MIDNIGHT TIME GONE BY CARDS

1 hour	2 hours	3 hours	4 hours	5 hours	6 hours
5 minutes	10 minutes	15 minutes	20 minutes	25 minutes	30 minutes
35 minutes	40 minutes	45 minutes	50 minutes	55 minutes	60 minutes
65 minutes	70 minutes	75 minutes	80 minutes	85 minutes	90 minutes
95 minutes	100 minutes	105 minutes	110 minutes	115 minutes	120 minutes
125 minutes	130 minutes	135 minutes	140 minutes	145 minutes	150 minutes

RACE TO MIDNIGHT RECORDING SHEET

Start Time	Time Gone By	End Time

TANGRAMS ANSWER KEY

Square

Butterfly

Trapezoid

Whale

COMPOSING HEXAGONS I

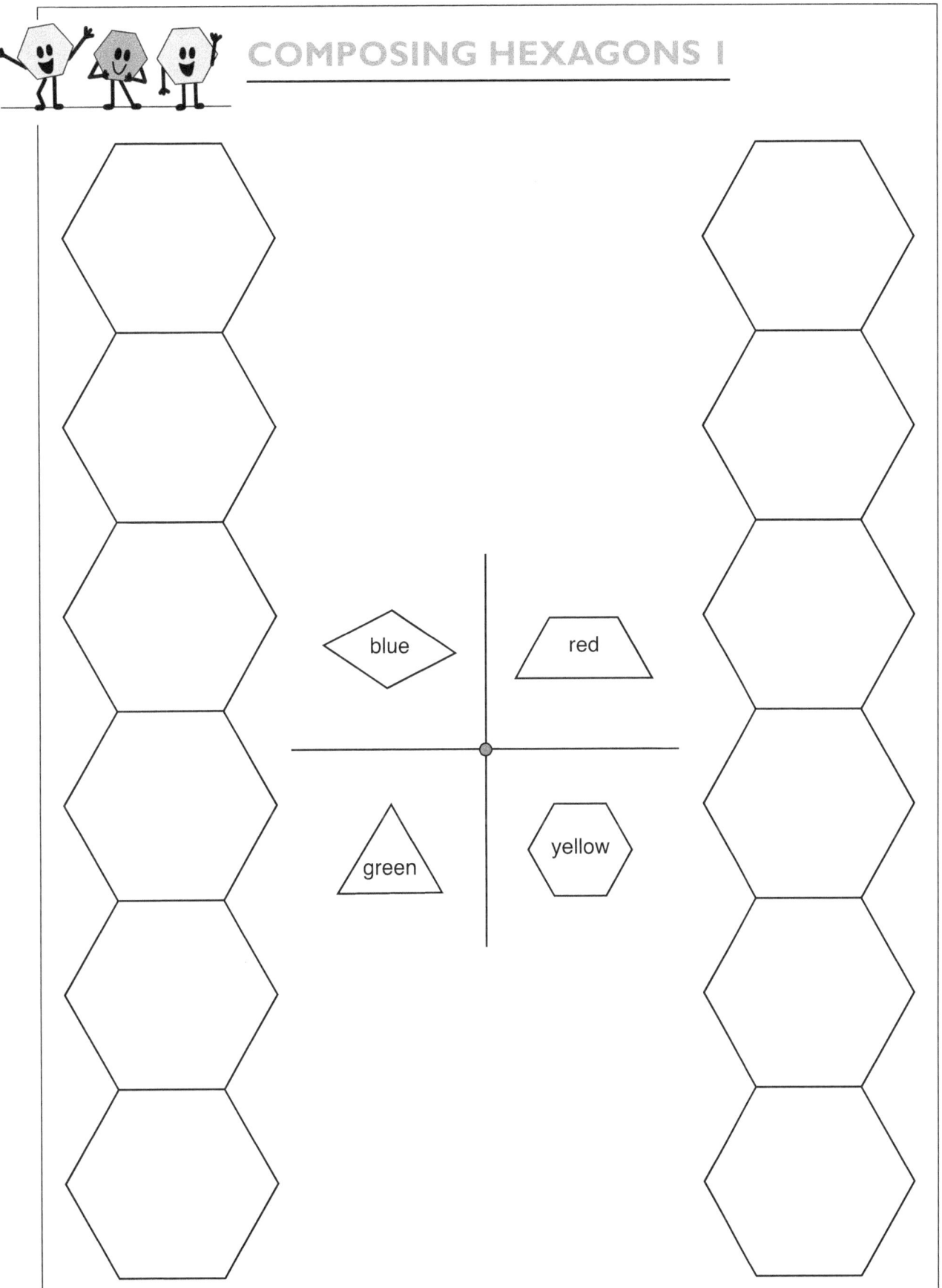

COMPOSING HEXAGONS 2

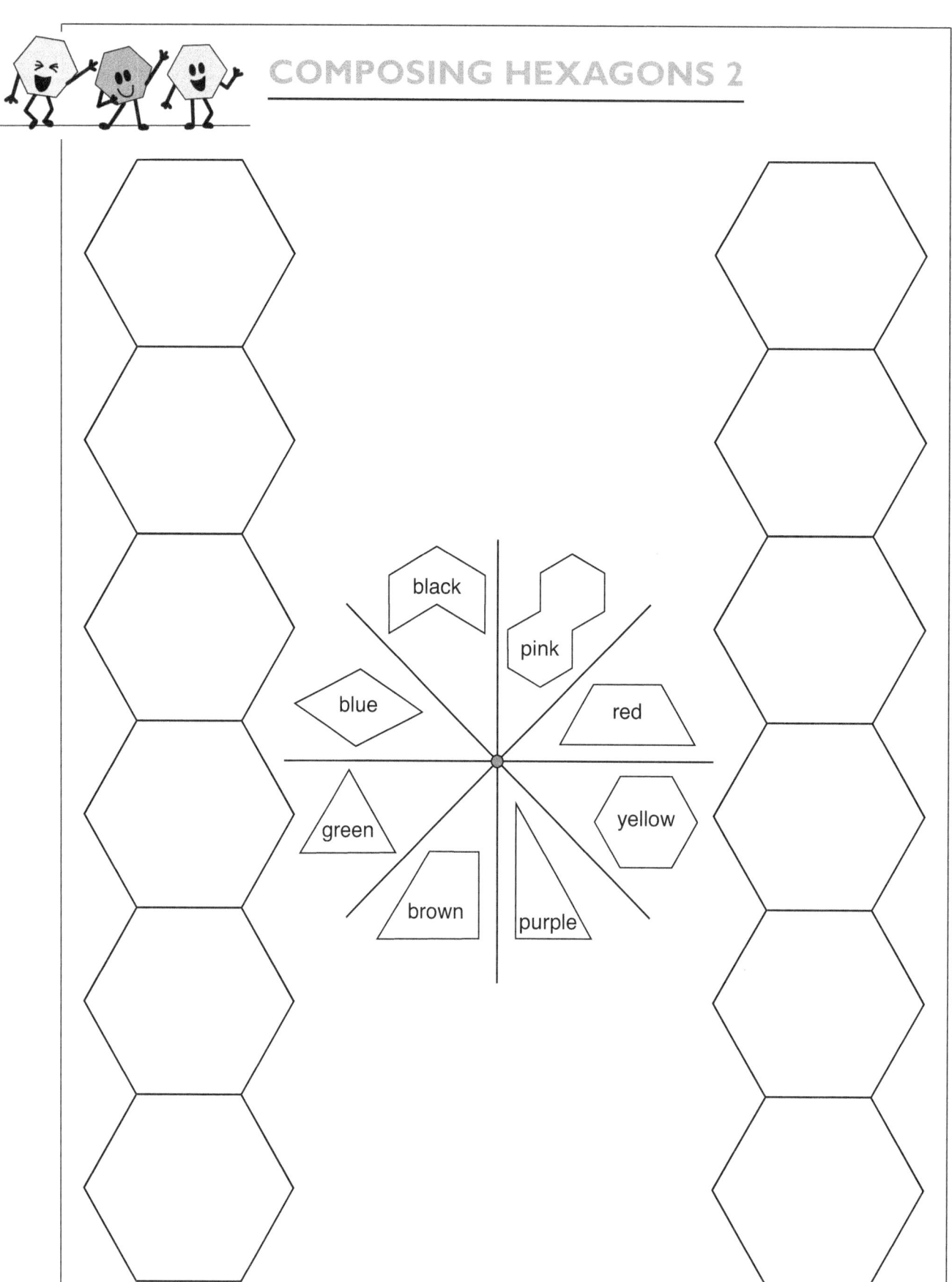

PATTERN BLOCKS NAME CHART

All Pattern Blocks represent polygons.

Some polygons have several names.

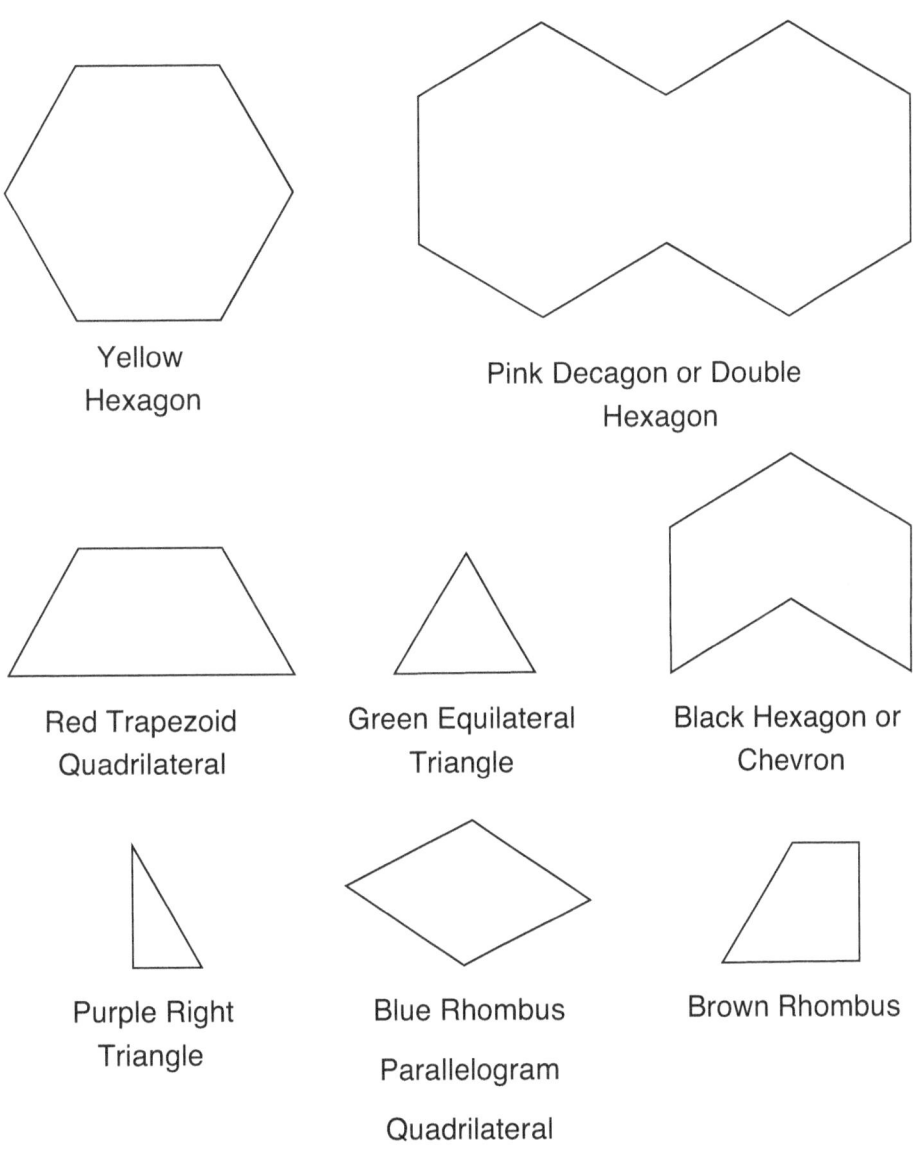

Yellow Hexagon

Pink Decagon or Double Hexagon

Red Trapezoid Quadrilateral

Green Equilateral Triangle

Black Hexagon or Chevron

Purple Right Triangle

Blue Rhombus Parallelogram Quadrilateral

Brown Rhombus

SIGMUND SQUARE QUESTION CARDS P-2

Question: When Sigmund changes color, is he still a square?	**Question:** How many sides and angles does a pentagon have?	**Question:** What is the name of the polygon that has six sides and six angles?	**Question:** Do all rectangles have two short sides and two long sides?
Answer: Yes	**Answer:** 5	**Answer:** Hexagon	**Answer:** No, some rectangles are squares with four equal sides
Move 1 square	**Move** 2 squares	**Move** 1 square	**Move** 2 squares
Question: Can you compose a square from two triangles?	**Question:** How many triangles do you need to compose a regular hexagon?	**Question:** If a square is turned onto its corner, is it called a diamond?	**Question:** Square + triangle + trapezoid = how many sides?
Answer: Yes	**Answer:** 6	**Answer:** No	**Answer:** 11
Move 1 square	**Move** 1 square	**Move** 1 square	**Move** 1 square
Question: How many more sides does a hexagon have than a rectangle?	**Question:** Do all polygons need to have sides of equal lengths?	**Question:** Are triangles quadrilaterals?	**Question:** Why is a square part of the rectangle family?
Answer: 2	**Answer:** No	**Answer:** No	**Answer:** A square is a quadrilateral with four right angles
Move 1 square	**Move** 1 square	**Move** 2 squares	**Move** 1 square
Question: How many more angles does a square have than a triangle?	**Question:** How many line segments do you need to draw 2 separate squares?	**Question:** How many line segments do you need to draw two squares connected on one full side?	**Question:** Are hexagons quadrilaterals?
Answer: 1	**Answer:** 8	**Answer:** 7	**Answer:** No
Move 1 square	**Move** 1 square	**Move** 1 square	**Move** 2 squares
Question: Is a circle a polygon?	**Question:** Do all polygons have straight sides?	**Question:** Square + rectangle + hexagon = how many sides?	**Question:** Can you compose a rectangle from four triangles?
Answer: No, polygons are closed shapes with all straight sides	**Answer:** Yes	**Answer:** 14	**Answer:** Yes
Move 2 squares	**Move** 1 square	**Move** 1 square	**Move** 1 square

SIGMUND SQUARE QUESTION CARDS 3–5

Question: Do you agree that Sigmund's evening walk "paralleled" his morning walk?	**Question:** How many sides and angles does a dodecagon have?	**Question:** What is the name of the polygon that has 10 sides and 10 angles?	**Question:** Which polygons have all right angles?
Answer: Yes	**Answer:** 12	**Answer:** Decagon	**Answer:** Square rectangles and non-square rectangles
Move 1 square	**Move** 2 squares	**Move** 1 square	**Move** 2 squares
Question: Do all rectangles have two short sides and two long sides?	**Question:** How many triangles do you need to compose a regular octagon?	**Question:** How is an acute angle different from a right angle?	**Question:** As more sides are added to a polygon, do the angles get larger or smaller?
Answer: No, some rectangles are squares with all equal sides	**Answer:** 8	**Answer:** Acute angle is less than 90°, right angle is exactly 90°	**Answer:** Larger
Move 2 squares	**Move** 1 square	**Move** 2 squares	**Move** 2 squares
Question: Square + heptagon + pentagon = how many sides?	**Question:** How many more sides does an icosagon have than a nonagon?	**Question:** Are trapezoids quadrilaterals?	**Question:** If a group of quadrilaterals had a total of 20 angles, how many quadrilaterals would there be?
Answer: 16	**Answer:** 11	**Answer:** Yes	**Answer:** 5
Move 1 square	**Move** 2 squares	**Move** 2 squares	**Move** 2 squares
Question: Do you agree that all squares are rectangles, but not all rectangles are squares?	**Question:** Why is a square part of the rectangle family?	**Question:** How many line segments are needed to draw 100 separate quadrilaterals?	**Question:** What makes a trapezoid different from a rectangle?
Answer: Yes	**Answer:** A square is a quadrilateral with four right angles	**Answer:** 400	**Answer:** A trapezoid is a quadrilateral with only one set of parallel sides
Move 1 square	**Move** 1 square	**Move** 1 square	**Move** 1 square
Question: Are rhombuses, rectangles, and squares considered quadrilaterals?	**Question:** What is a right triangle?	**Question:** Do all polygons have angles?	**Question:** Which shape is not a quadrilateral: rectangle, square, rhombus, hexagon, parallelogram?
Answer: Yes	**Answer:** A right triangle is a three-sided polygon with one right angle (90°)	**Answer:** Yes	**Answer:** Hexagon
Move 1 square	**Move** 2 squares	**Move** 1 square	**Move** 1 square

SIGMUND SQUARE GAME BOARD

Sigmund Square Finds His Family

SNACK SHOP CARDS

The 🌭 costs $ ____ .
The 🍿 costs ____ times more than the 🌭 .
How much does the 🍿 cost?

The 🍟 cost $ ____ .
The 🍔 costs ____ times more than the 🍟 .
How much does the 🍔 cost?

The 🍦 costs $ ____ .
The 🍕 costs ____ times more than the 🍦 .
How much does the 🍕 cost?

The 🍿 costs $ ____ .
The 🥪 costs ____ times more than the 🍿 .
How much does the 🥪 cost?

The 🥤 costs $ ____ .
The ☕ costs ____ times more than the 🥤 .
How much does the ☕ cost?

FLIP TOP DIVISION BOTTLE CAP LABELS

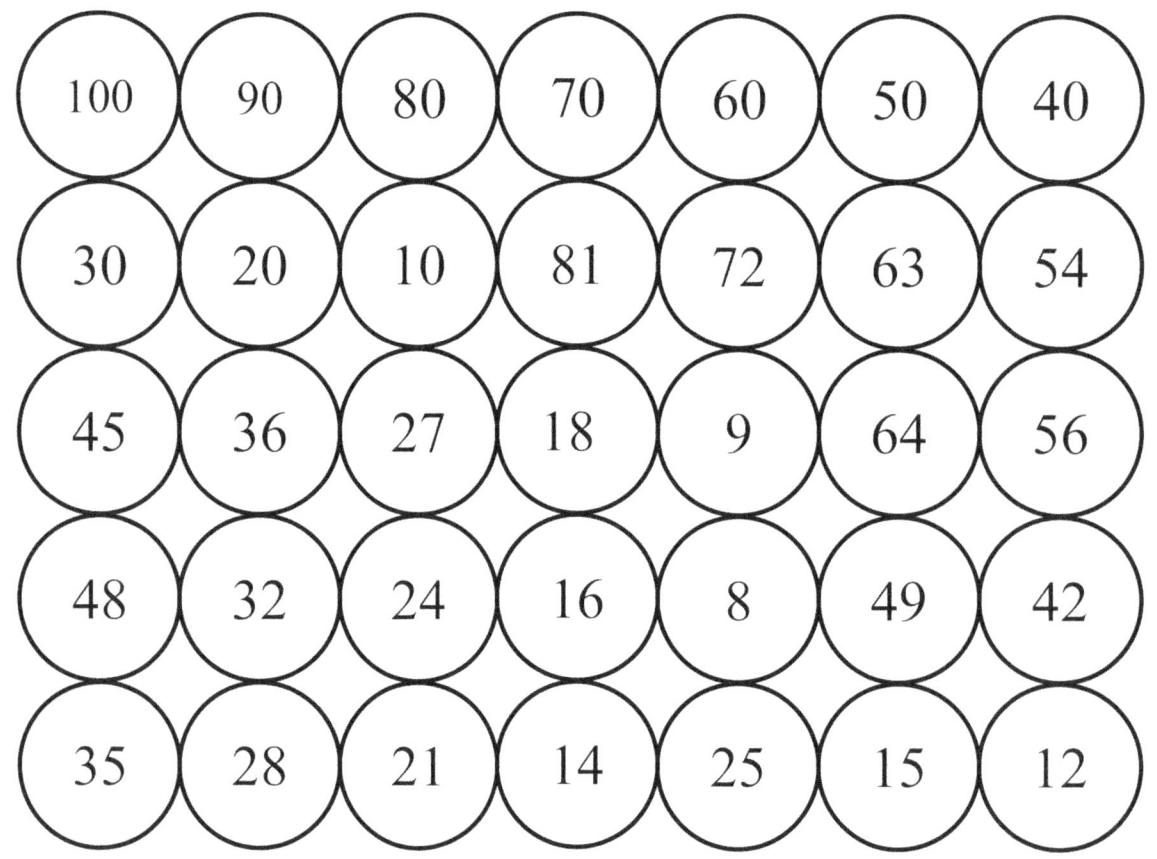

PLACE THE DIGIT GAME BOARDS

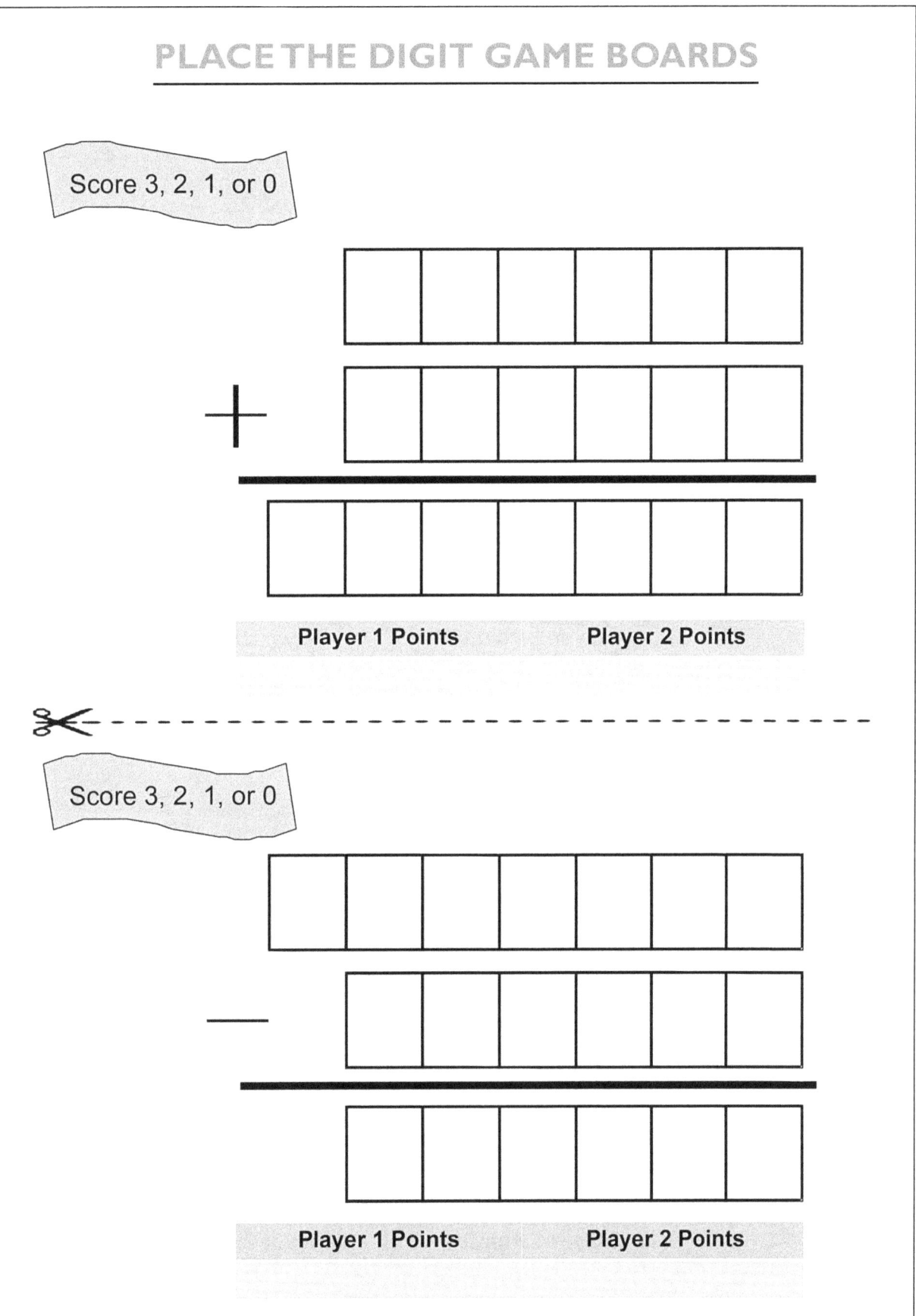

PLACE THE DIGIT GAME BOARDS

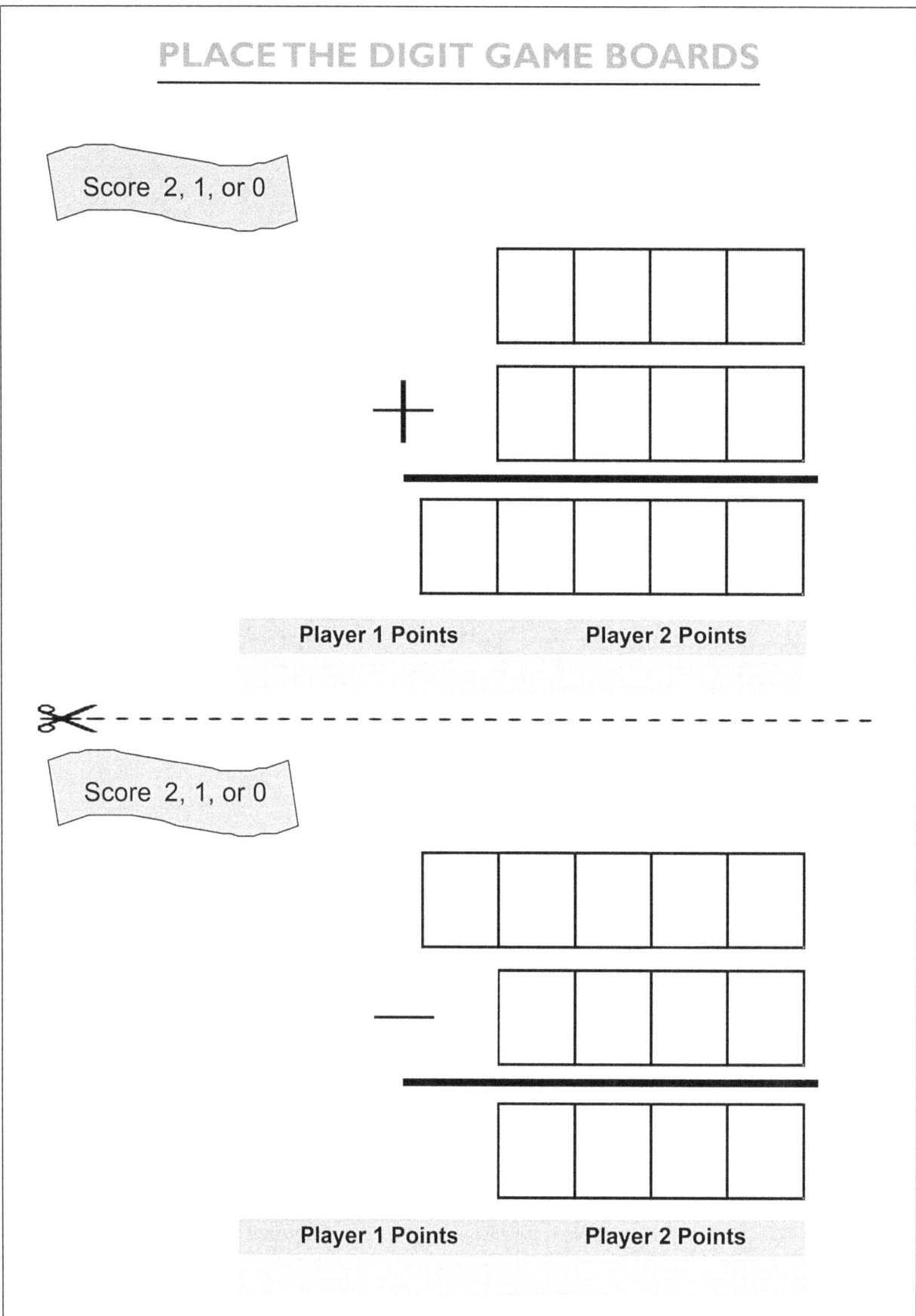

PLACE THE DIGIT GAME BOARDS

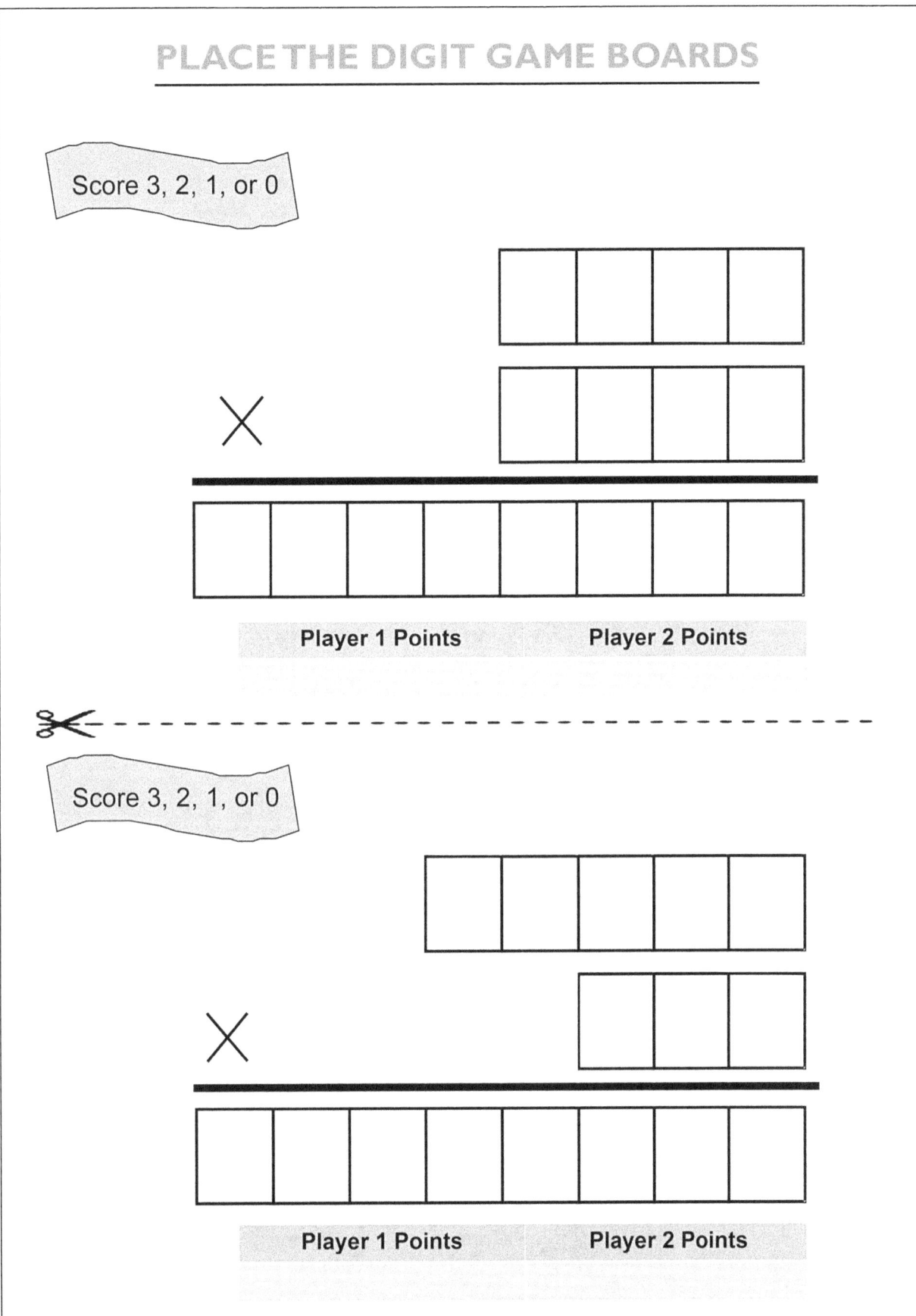

PLACE THE DIGIT GAME BOARDS

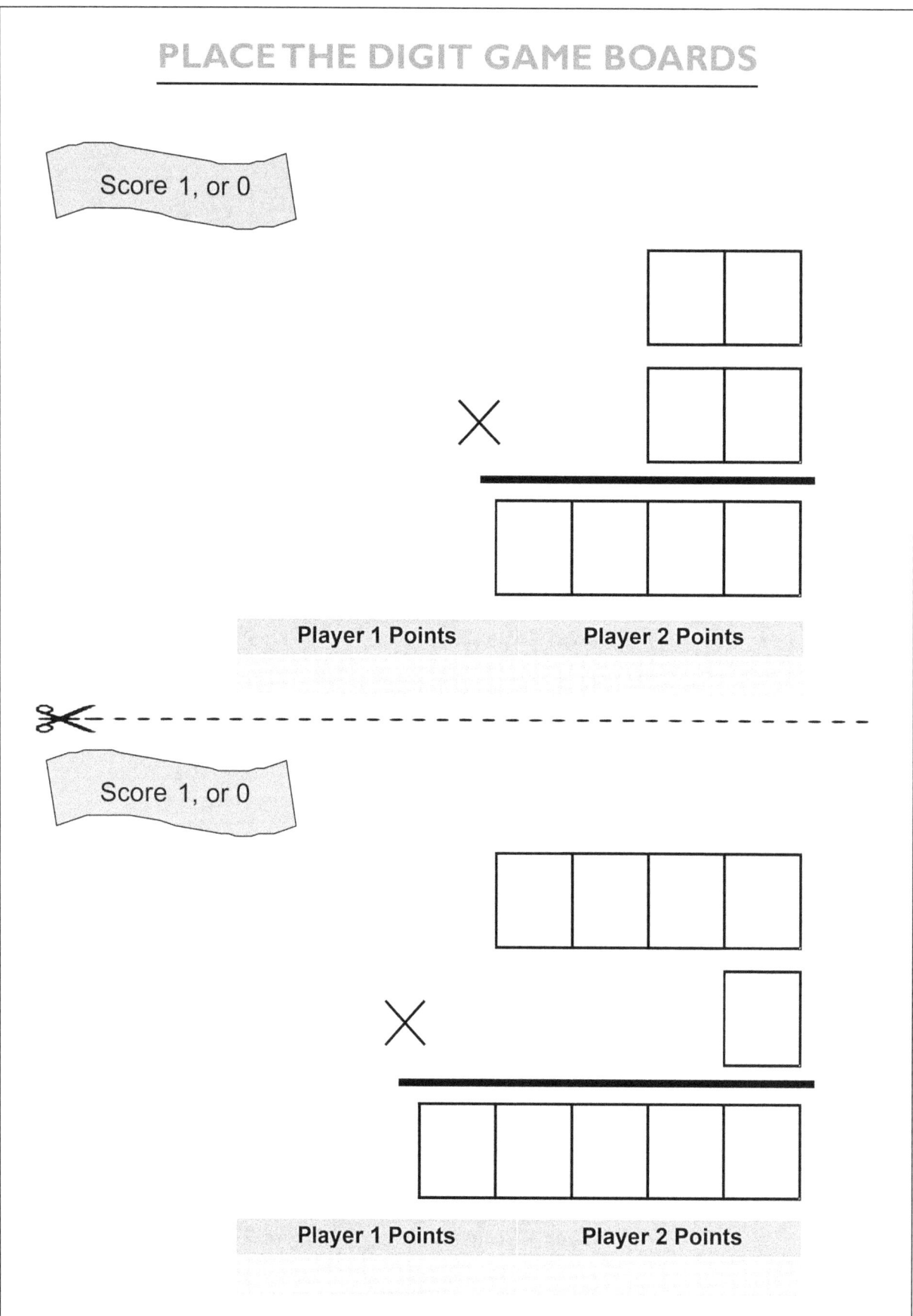

CARD MATH GAME BOARD

HEADS UP DECIMAL CARDS

One tenth	1	Eight hundredths	One and one tenth	$\frac{7}{10}$	.07	.5
$\frac{12}{100}$	.04	Nine tenths	.1	Six hundredths	One hundredth	.08
Five tenths	.11	.06	Ten tenths	One	$\frac{3}{10}$	.03
Ten hundredths	.4	1.02	Nine hundredths	.7	1.2	Six tenths
Four hundredths	.05	Two tenths	1.1	.9	$1\frac{3}{100}$	.8
.01	$\frac{8}{10}$	Eleven hundredths	.6	Four tenths	.09	$\frac{5}{100}$
$\frac{10}{10}$	.3	Seven hundredths	.02	.12	Three hundredths	.2
1.01	One and two tenths	One and two hundredths	1.0	$\frac{2}{100}$	One and one hundredth	1.03

HEADS UP DECIMAL GRID

Heads Up Decimal Grid   one hundredth   one tenth

WHERE'S $\frac{1}{2}$? NUMBER LINES

WHERE'S $\frac{1}{2}$? NUMBER LINES

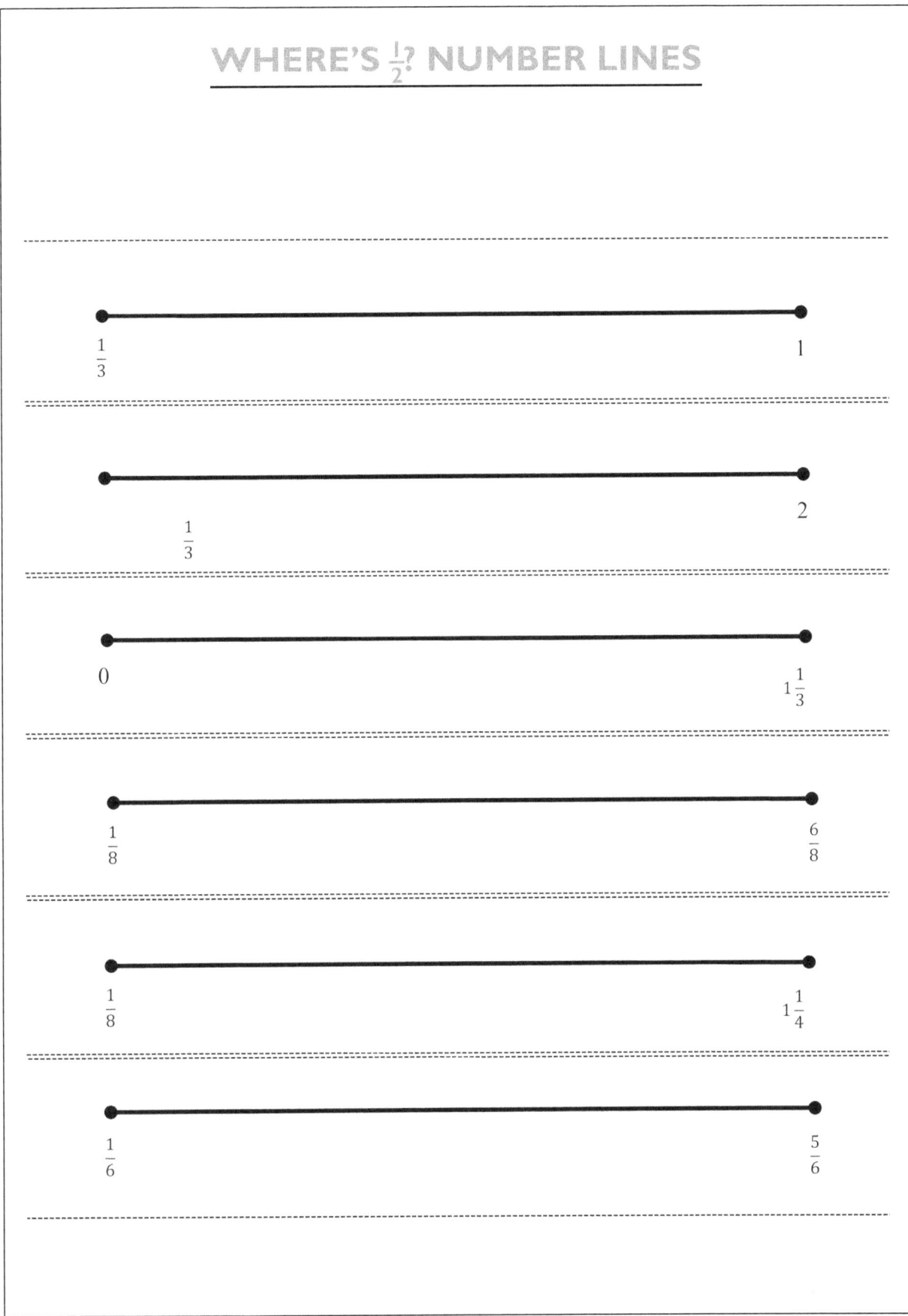

WHERE'S $\frac{1}{2}$? NUMBER LINES

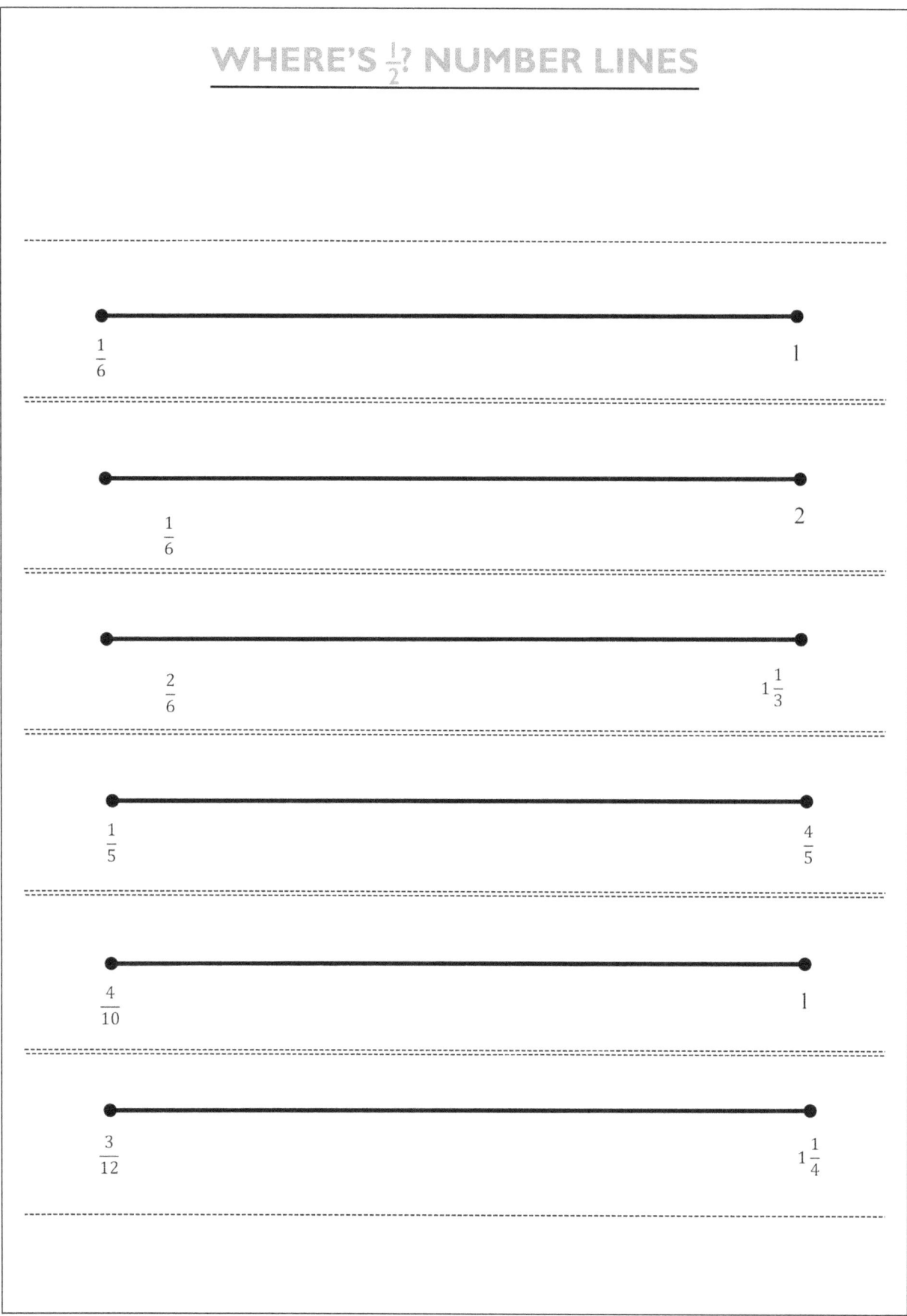

FRACTION FISH

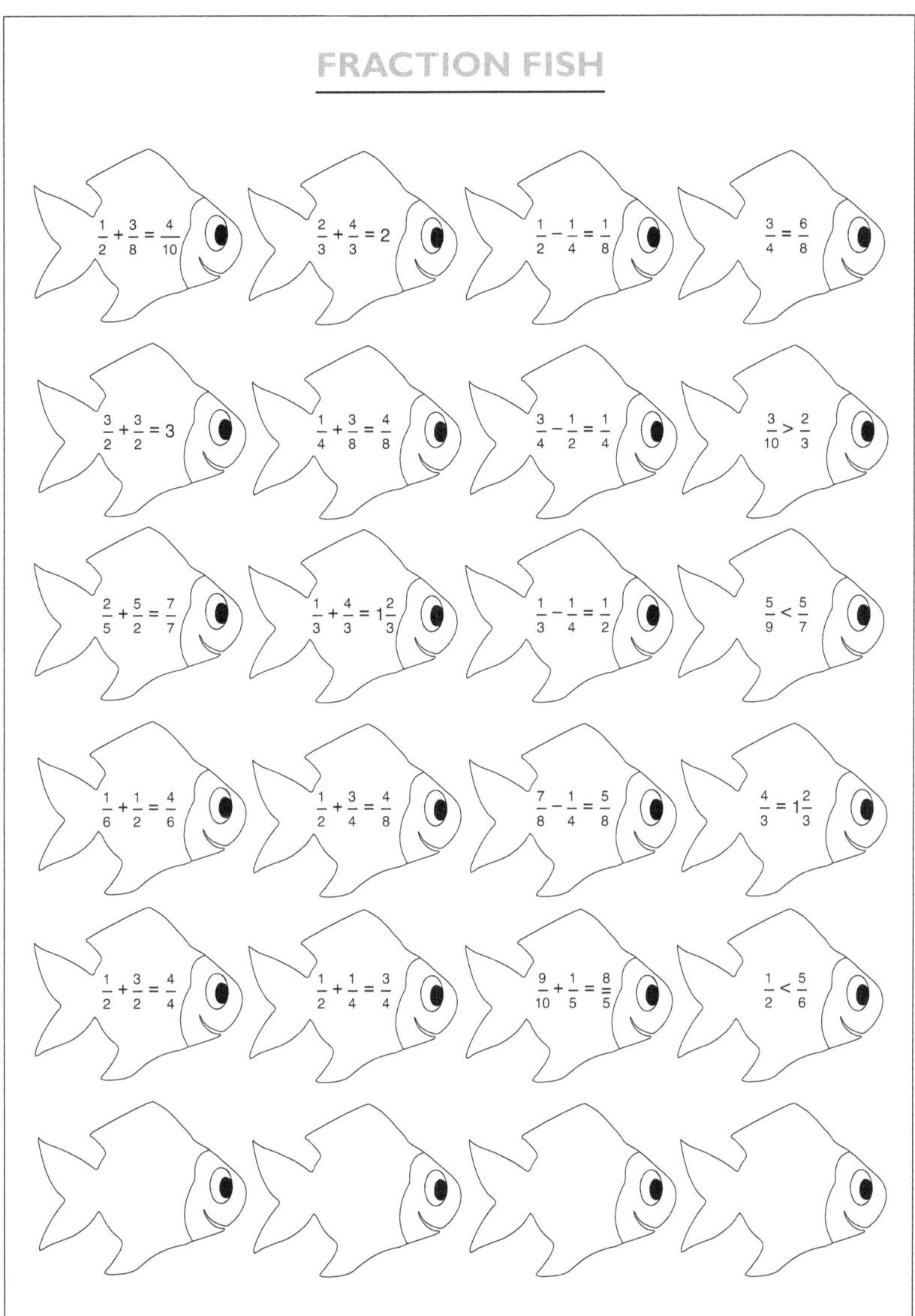

Row 1: $\frac{1}{2}+\frac{3}{8}=\frac{4}{10}$; $\frac{2}{3}+\frac{4}{3}=2$; $\frac{1}{2}-\frac{1}{4}=\frac{1}{8}$; $\frac{3}{4}=\frac{6}{8}$

Row 2: $\frac{3}{2}+\frac{3}{2}=3$; $\frac{1}{4}+\frac{3}{8}=\frac{4}{8}$; $\frac{3}{4}-\frac{1}{2}=\frac{1}{4}$; $\frac{3}{10}>\frac{2}{3}$

Row 3: $\frac{2}{5}+\frac{5}{2}=\frac{7}{7}$; $\frac{1}{3}+\frac{4}{3}=1\frac{2}{3}$; $\frac{1}{3}-\frac{1}{4}=\frac{1}{2}$; $\frac{5}{9}<\frac{5}{7}$

Row 4: $\frac{1}{6}+\frac{1}{2}=\frac{4}{6}$; $\frac{1}{2}+\frac{3}{4}=\frac{4}{8}$; $\frac{7}{8}-\frac{1}{4}=\frac{5}{8}$; $\frac{4}{3}=1\frac{2}{3}$

Row 5: $\frac{1}{2}+\frac{3}{2}=\frac{4}{4}$; $\frac{1}{2}+\frac{1}{4}=\frac{3}{4}$; $\frac{9}{10}+\frac{1}{5}=\frac{8}{5}$; $\frac{1}{2}<\frac{5}{6}$

BEAR SLIDE RESULTS CARDS

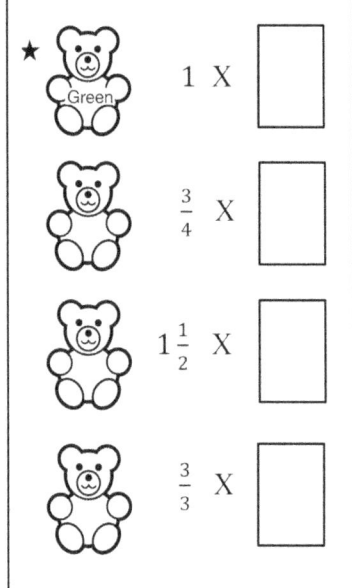
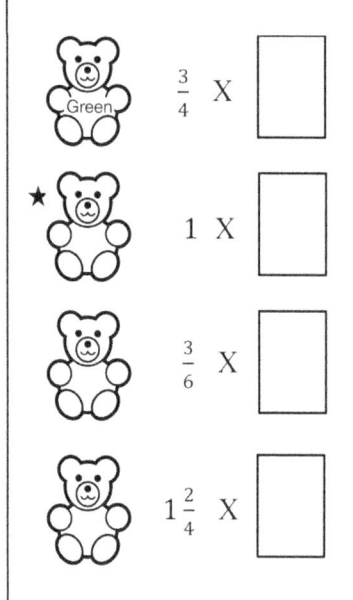
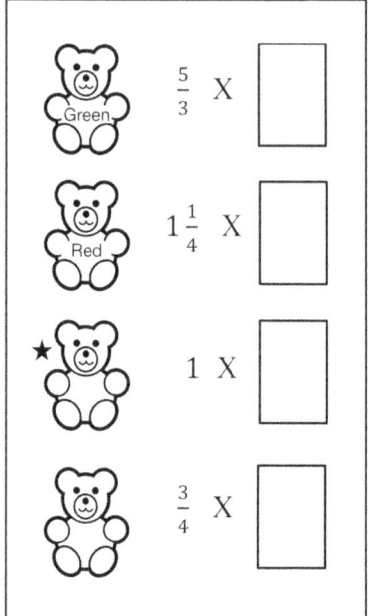
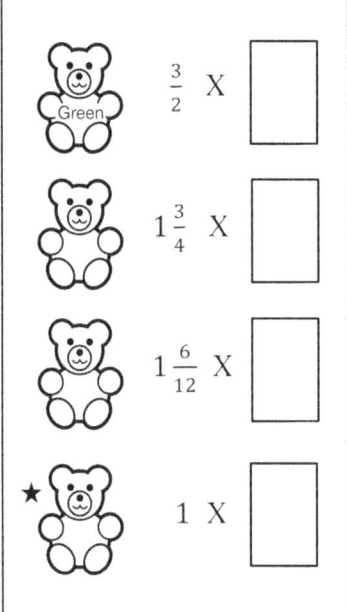
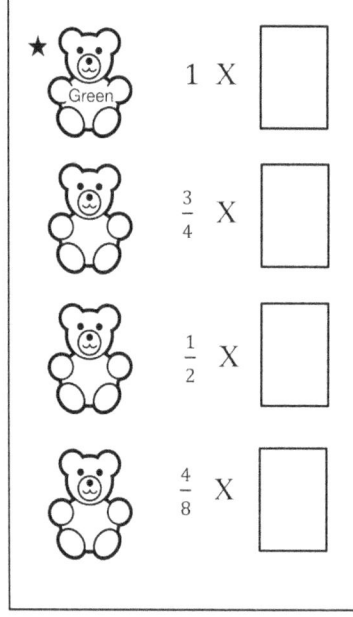
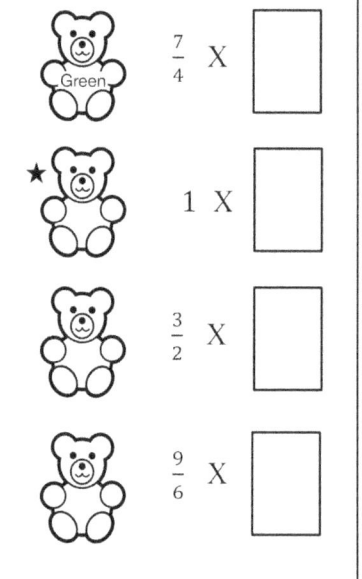

BEAR SLIDE NUMBER LINES

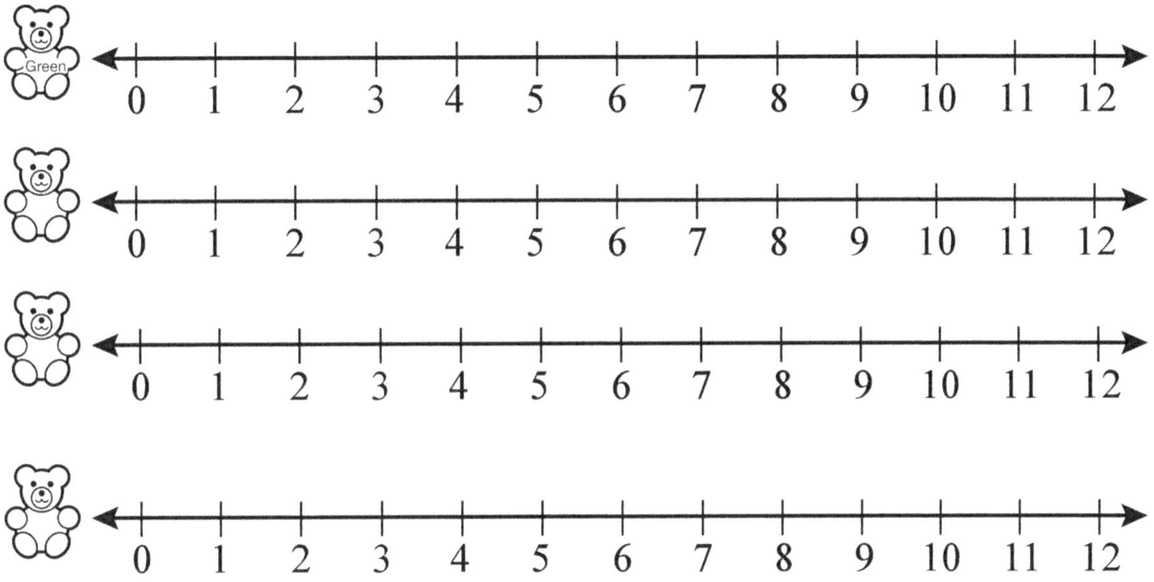

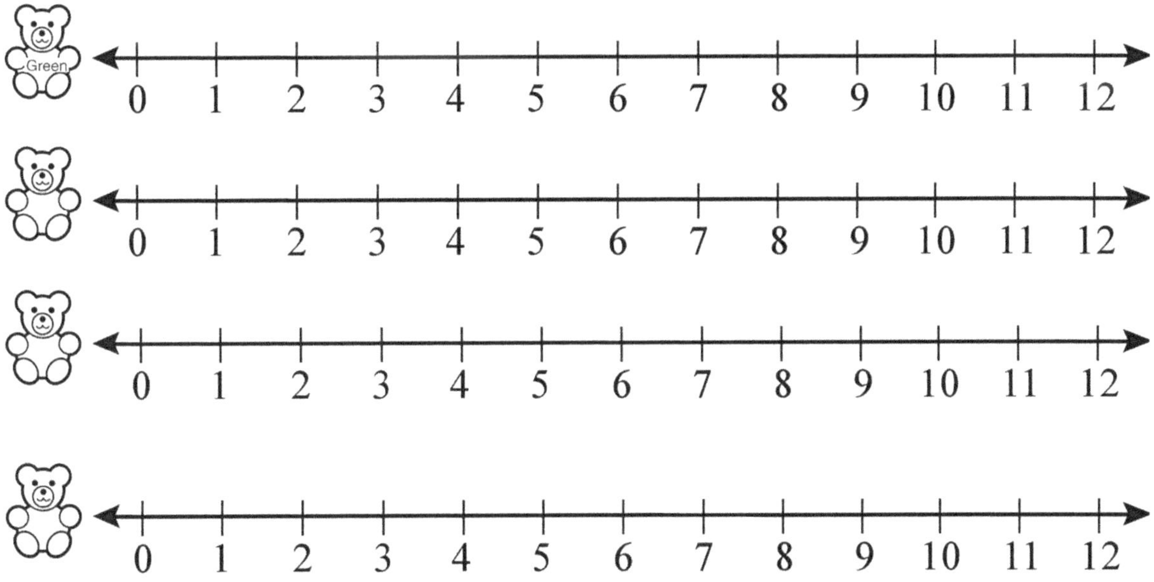

CATERPILLARS

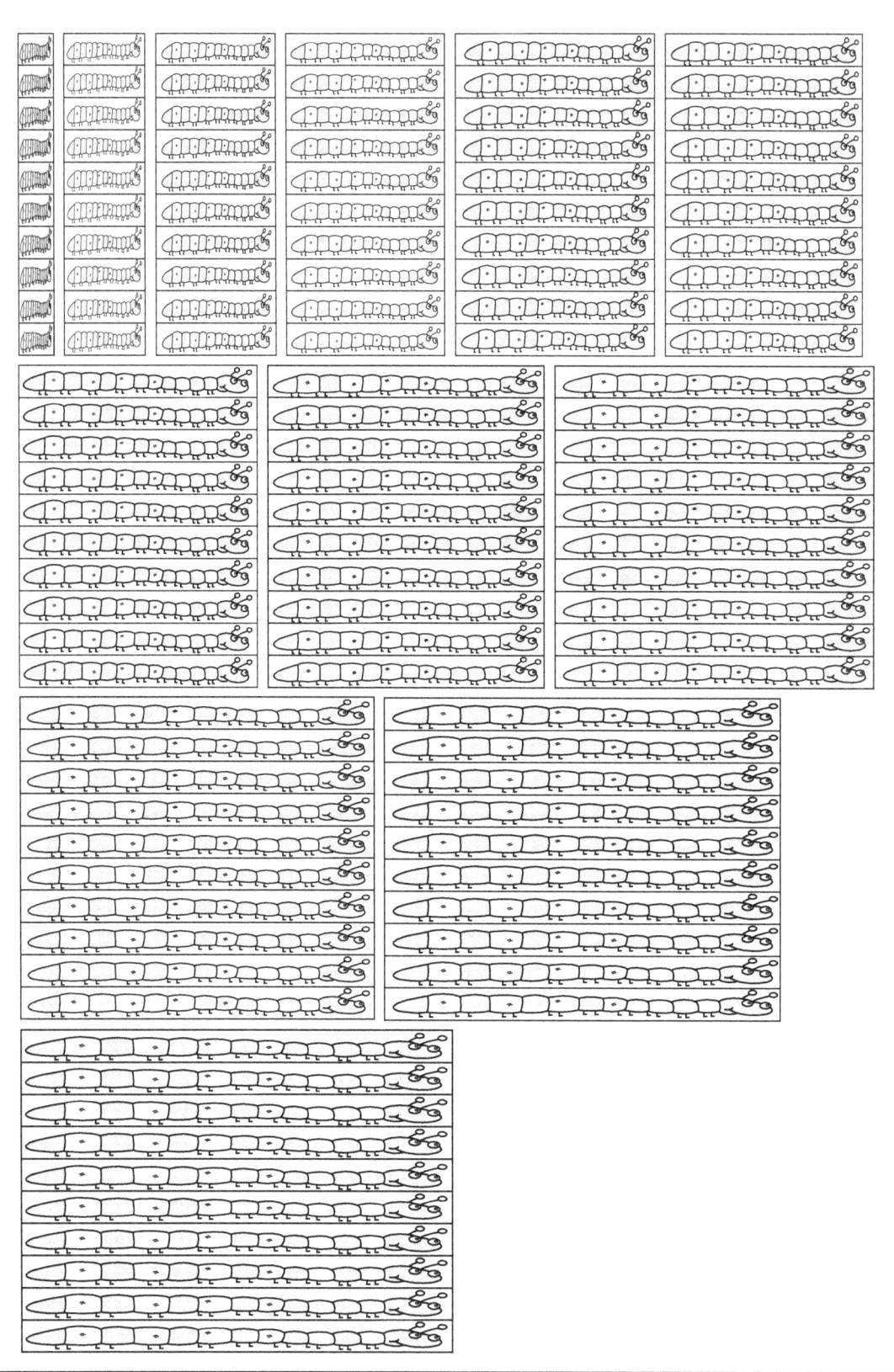

ANGLE FACE CARDS

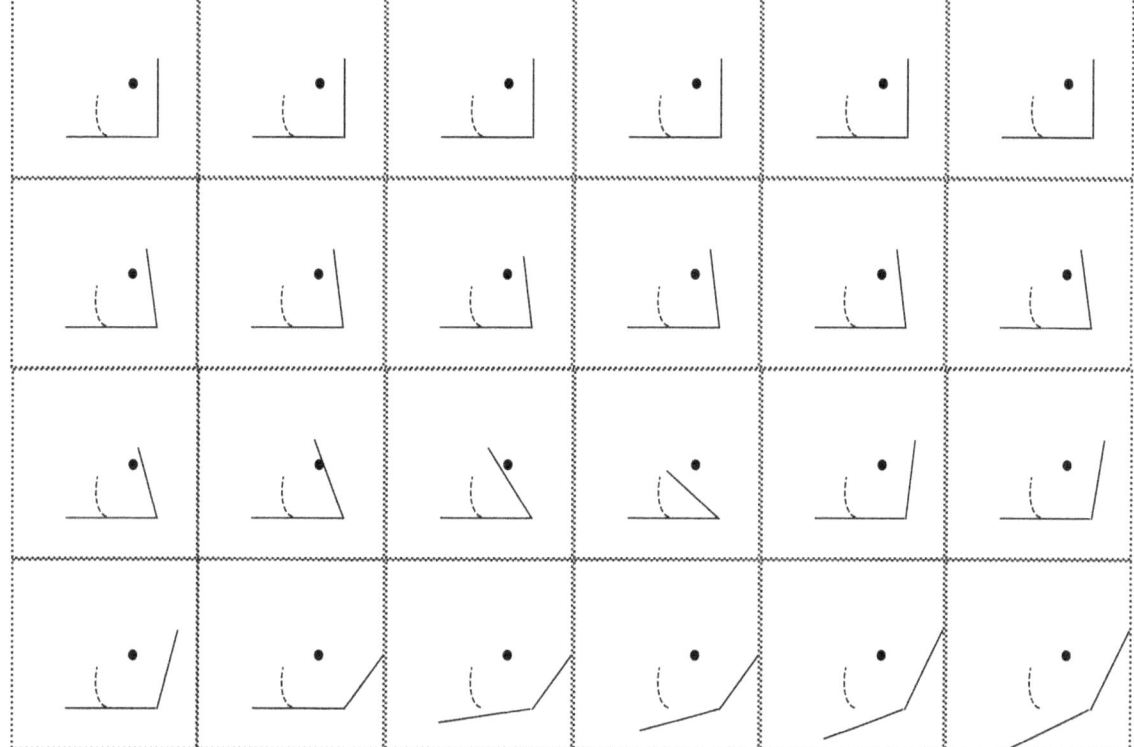

© 2017, *Family Math Night*, Jennifer Taylor-Cox, Routledge

ANGLE FACE SORTING MATS

Acute Angle	Right Angle	Obtuse Angle

DRAW WHAT I SAY CARDS

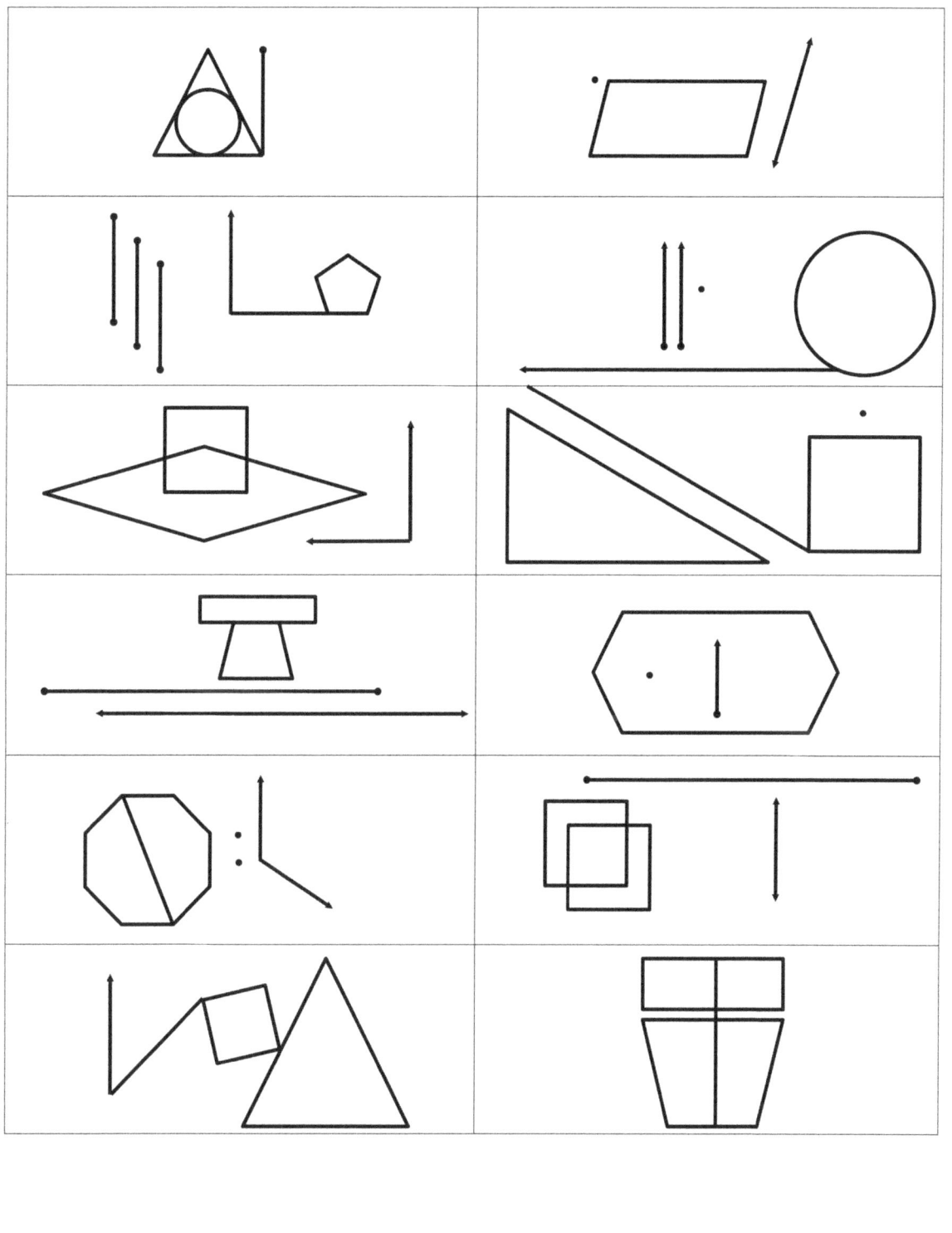

PENTOMINOES

Pentominoes: Answer Key

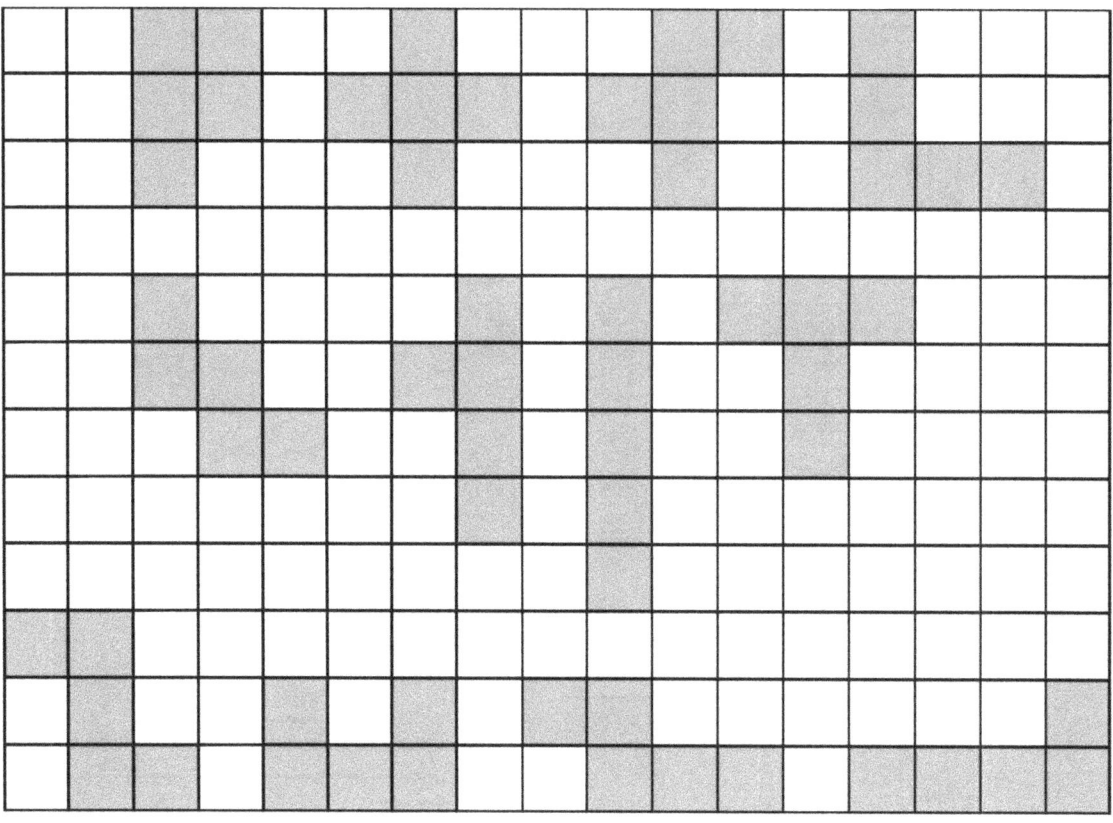

THREE SMILES COORDINATE PLANE

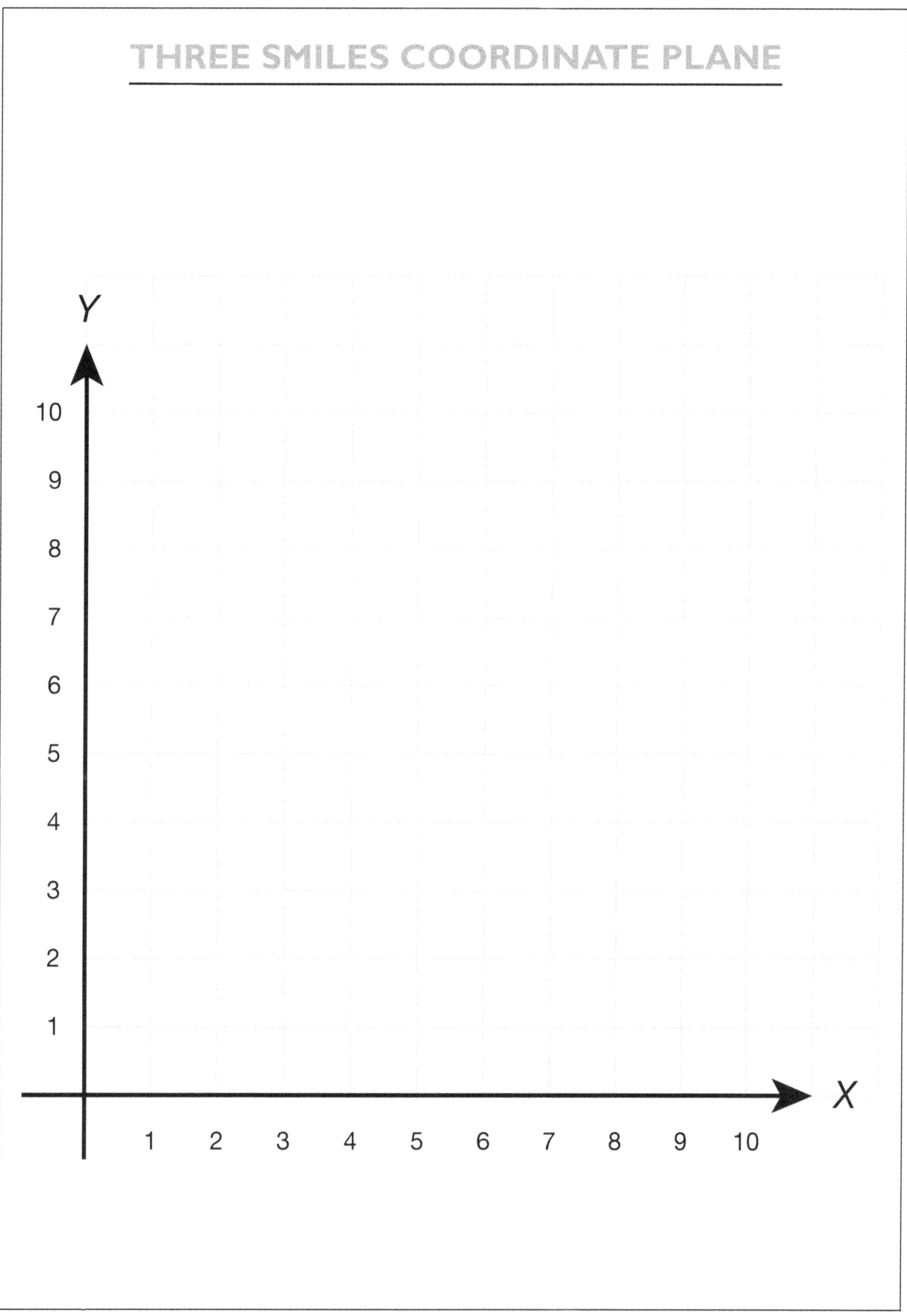

GEOMETRY VOCABULARY

Acute: An acute angle is a positive angle that is less than 90°. An acute triangle has three angles that are each less than 90°.

Angle: An angle is the endpoint shared by two rays or line segments. The corner or point where the lines meet and end can also be called a vertex (plural vertices). When two lines meet at a vertex, an angle is formed. The number and size of the interior angles help define polygons.

Coordinates: Coordinates are an ordered pair of numbers that locate an exact position on a coordinate grid/plane.

Decagon: A decagon is a polygon with 10 sides and 10 angles. The sum of the interior angles of a decagon is 1,440°. The general formula is $180(n-2)$.

Dilation: A dilation of a polygon is a transformation of the shape where the size is larger (enlargement/stretch) or smaller (reduction/shrink) using a scale factor (ratio) and the center of the dilation.

Dodecagon: A dodecagon is a polygon with 12 sides and 12 angles. The sum of the interior angles of a dodecagon is 1,800°. The general formula is $180(n-2)$.

Equal: Equal means having the same value, quantity, or measure.

Equilateral triangle: An equilateral triangle has all sides that are the same length (congruent sides).

Flip: A flip (or reflection) of a polygon is a transformation of the shape where each point is at equal distance on the opposite side of a given line (called the line of reflection).

Hendecagon: A hendecagon (or undecagon) is a polygon with 11 sides and 11 angles. The sum of the interior angles of a hendecagon is 1,620°. The general formula is $180(n-2)$.

GEOMETRY VOCABULARY

Heptagon: A heptagon is a polygon with seven sides and seven angles. The sum of the interior angles of a heptagon is 900°. The general formula is $180(n-2)$.

Hexagon: A hexagon is a polygon with six sides and six angles. The sum of the interior angles of a hexagon is 720°. The general formula is $180(n-2)$.

Irregular: Irregular polygons have sides and angles that are not equal. A polygon is regular if all sides and all angles are equal.

Kite: A kite is a quadrilateral with two distinct pairs of equal adjacent sides.

Nonagon: A nonagon (or enneagon) is a polygon with nine sides and nine angles. The sum of the interior angles of a nonagon is 1,260°. The general formula is $180(n-2)$.

Obtuse: An obtuse angle is more than 90° and less than 180°. An obtuse triangle has one angle that is greater than 90°.

Octagon: An octagon is a polygon with eight sides and eight angles. The sum of the interior angles of an octagon is 1,080°. The general formula is $180(n-2)$.

Parallel: Lines are parallel when they lie in the same plane and are always the same distance apart. Perpendicular lines are the opposite of parallel lines because they intersect at right angles.

Parallelogram: A parallelogram is a quadrilateral with two sets of parallel sides.

Pentagon: A pentagon is a polygon with five sides and five angles. The sum of the interior angles of a pentagon is 540°. The general formula is $180(n-2)$.

Perimeter: The perimeter is the distance around the outside of a shape.

GEOMETRY VOCABULARY

Polygon: A polygon is a closed shape with all straight sides. Regular polygons are equiangular (all angles equal) and equilateral (all sides equal).

Quadrilateral: A quadrilateral is a polygon with four sides and four angles.

Rectangle: A rectangle is a quadrilateral with four right angles. A rectangle has four interior angles that are each 90°. The sum of the interior angles is 360°.

Reflection: A reflection (or flip) of a polygon is a transformation of the shape where each point is at equal distance on the opposite side of a given line (called the line of reflection).

Rhombus: A rhombus is a parallelogram with all sides the same length (congruent sides).

Right angle: A right angle is an angle that measures 90°.

Scalene triangle: A scalene triangle is a triangle with three sides that are different lengths.

Side: Polygons are made of line segments. Each line segment is considered a side.

Similarity: In math, similar figures are the same shape. The corresponding angles are equal (congruent) and the lengths of their corresponding sides have the same ratio.

Slide: A slide (or translation) of a polygon is a transformation of the shape where the shape moves along a given line.

Square: A square is a quadrilateral with four right angles (rectangle) and all sides equal in length. It is considered a special type of rectangle. A square has four interior angles that are each 90°. The sum of the interior angles of a square is 360°.

GEOMETRY VOCABULARY

Square root: In math, a square root is a number that when multiplied by itself produces a given number. For example, the square root of 16 (written $\sqrt{16}$) is 4 because 4 x 4 = 16.

Transformation: A congruent transformation is a change in position. Flips, slides, and turns are types of congruent transformations in plane symmetry. A dilation is a transformation that is not congruent, but is similar.

Trapezoid: A trapezoid is a quadrilateral with only one set of parallel sides.

Triangle: A triangle is a polygon with three sides and three angles. The sum of the interior angles of a triangle is 180°. Types of triangles are named by sides or angles.

Turn: A turn (or rotation) of a polygon is a transformation of the shape where the shape rotates on a point (called the center of rotation).

Helping you to choose the right eBooks for your Library

Add Routledge titles to your library's digital collection today. Taylor and Francis ebooks contains over 50,000 titles in the Humanities, Social Sciences, Behavioural Sciences, Built Environment and Law.

Choose from a range of subject packages or create your own!

Benefits for you
- Free MARC records
- COUNTER-compliant usage statistics
- Flexible purchase and pricing options
- All titles DRM-free.

Benefits for your user
- Off-site, anytime access via Athens or referring URL
- Print or copy pages or chapters
- Full content search
- Bookmark, highlight and annotate text
- Access to thousands of pages of quality research at the click of a button.

 **REQUEST YOUR FREE INSTITUTIONAL TRIAL TODAY**

Free Trials Available
We offer free trials to qualifying academic, corporate and government customers.

eCollections – Choose from over 30 subject eCollections, including:

Archaeology	Language Learning
Architecture	Law
Asian Studies	Literature
Business & Management	Media & Communication
Classical Studies	Middle East Studies
Construction	Music
Creative & Media Arts	Philosophy
Criminology & Criminal Justice	Planning
Economics	Politics
Education	Psychology & Mental Health
Energy	Religion
Engineering	Security
English Language & Linguistics	Social Work
Environment & Sustainability	Sociology
Geography	Sport
Health Studies	Theatre & Performance
History	Tourism, Hospitality & Events

For more information, pricing enquiries or to order a free trial, please contact your local sales team:
www.tandfebooks.com/page/sales

 **Routledge** — Taylor & Francis Group | The home of Routledge books

www.tandfebooks.com

For Product Safety Concerns and Information please contact our EU
representative GPSR@taylorandfrancis.com
Taylor & Francis Verlag GmbH, Kaufingerstraße 24, 80331 München, Germany

www.ingramcontent.com/pod-product-compliance
Lightning Source LLC
Chambersburg PA
CBHW081421230426
43668CB00016B/2311